# Lecture Notes in Mobility

The book series Lecture Notes in Mobility (LNMOB) reports on innovative, peer-reviewed research and developments in intelligent, connected and sustainable transportation systems of the future. It covers technological advances, research, developments and applications, as well as business models, management systems and policy implementation relating to: zero-emission, electric and energy-efficient vehicles; alternative and optimized powertrains; vehicle automation and cooperation; clean, user-centric and on-demand transport systems; shared mobility services and intermodal hubs; energy, data and communication infrastructure for transportation; and micromobility and soft urban modes, among other topics. The series gives a special emphasis to sustainable, seamless and inclusive transformation strategies and covers both traditional and any new transportation modes for passengers and goods. Cutting-edge findings from public research funding programs in Europe, America and Asia do represent an important source of content for this series. PhD thesis of exceptional value may also be considered for publication. Supervised by a scientific advisory board of world-leading scholars and professionals, the Lecture Notes in Mobility are intended to offer an authoritative and comprehensive source of information on the latest transportation technology and mobility trends to an audience of researchers, practitioners, policymakers, and advanced-level students, and a multidisciplinary platform fostering the exchange of ideas and collaboration between the different groups.

Henriette Cornet · Maria Gkemou
Editors

# Shared Mobility Revolution

Pioneering Autonomous Horizons

 Springer

*Editors*
Henriette Cornet
University of San Francisco
San Francisco, CA, USA

Maria Gkemou
Hellenic Institute of Transport
Centre for Research and Technology Hellas
(CERTH)
Athens, Greece

ISSN 2196-5544     ISSN 2196-5552 (electronic)
Lecture Notes in Mobility
ISBN 978-3-031-71792-5     ISBN 978-3-031-71793-2 (eBook)
https://doi.org/10.1007/978-3-031-71793-2

This Springer imprint is published by the registered company Springer Nature Switzerland AG
The registered company address is: Gewerbestrasse 11, 6330 Cham, Switzerland

If disposing of this product, please recycle the paper.

# Preface

In recent years, the landscape of transportation has been rapidly evolving, driven by advancements in technology, changing consumer preferences, and the imperative to address pressing environmental concerns. Among these changes, the emergence of automated mobility stands out as a transformative force, promising to revolutionize how people and goods move within cities and across regions.

This book, *Shared Mobility Revolution: Pioneering Autonomous Horizons*, is the initiative of the Horizon 2020 funded project SHOW, 'SHared automation Operating models for Worldwide adoption' (GA No 857730), which was at its start in 2020 the largest EU-initiative on large-scale deployment of automated shared mobility gathering a uniquely vast consortium of 70 European partners from industry, academia, and non-profit entities. SHOW aimed to advance sustainable urban transport through technical solutions, business models, and priority scenarios for impact assessment, by deploying shared, connected, electrified fleets of automated vehicles (AVs) in coordinated Public Transport (PT), Demand Responsive Transport (DRT), Mobility as a Service (MaaS) and Logistics as a Service (LaaS) in urban pilots all across Europe. These real-life urban pilots took place in 22 cities in Europe, with operations open to the public at each pilot site lasting between 9 and 12 months.

As editors of this series, we are pleased to present a collection of chapters that showcase some of the latest research findings from these real-world pilots and case studies from SHOW and beyond. Indeed, contributions from SHOW's sister projects like AUGMENTED CCAM[1] and FAME[2] highlights additional aspects of automated shared mobility and the role of collaboration with stakeholders in bridging it. Furthermore, several chapters explain the theoretical foundations and practical application of automated vehicles and provide further examples from pilot sites written by operators, experts, and practitioners.

The chapters in this series are organized into four distinct parts, each focusing on a key aspect of automated mobility.

---

[1] https://augmentedccam.com/.

[2] https://www.connectedautomateddriving.eu.

Part I, *Foundations of Automated Mobility*, lays the groundwork by examining the technological advancements underpinning shared automated mobility services, remote supervision strategies, and inclusive architecture frameworks. In the first chapter, Yun-Pang Flötteröd et al. present novel vehicle technologies for shared automated mobility services from real-life pilots, namely three types of automated vehicles, i.e., a U-Shift, a modular vehicle with an interchangeable capsule, a free-moving FZI-shuttle and a VIF-robotaxi, deployed at three sites in Germany and Austria. In the second chapter, Henriette Cornet et al. gather insights from first-hand practitioners, i.e., public transport operators involved in SHOW and other initiatives around the challenges and the requirements linked with remote supervision of automated vehicle fleets. In the third chapter, Anastasia Bolovinou et al. propose high-level framework and specifications for design alternatives and cloud-to-everything interfaces for automated vehicles deployments with an example of an inclusive architecture instantiated in Trikala, Greece.

In part II, *Infrastructure and Operations*, we delve into the critical role of physical and digital infrastructure in supporting automated vehicles, while also providing valuable insights gleaned from real-life operational experiences at various sites from SHOW including a bus depot and the deployment of automated logistic services. More specifically, the chapter from Andreas Hula et al. explores adaptations of the physical infrastructure with digital support to enable automated shared mobility services. In the second chapter of this part, focusing on three German SHOW sites, Yun-Pang Flötteröd et al. present fixed and non-fixed mobility services and the technical possibilities to let shuttles move without a pre-defined virtual track and cope with a ride-booking application. César Omar Chacón Fernández et al. delve into automated bus depot management. They present the EMT automated bus depot pilot at Carabanchel, Spain, and how it has been equipped with perception sensors, control mechanisms, and centralized decision-making units, testing services like internal transport, autoparking, and teleoperation to manage and monitor efficiently and safely a fleet of automated buses. The authors also present some first cost estimates of the operation from the operator's perspective. In the concluding chapter of part II, Venkata Akhil Babu Malisetty et al. expose a real-life logistics service that was operated with a fleet of five droids on a public route in the urban pedestrian area in Trikala, Greece. They describe the solutions implemented and the challenges faced by the droids while performing last-mile deliveries and provide recommendations for future implementations.

The part III, *User Perspectives and Engagement*, shifts the focus to the human factor, examining user acceptance, societal impacts, and stakeholder engagement in the context of shared automated mobility. Understanding the needs and preferences of users is essential for the successful implementation of automated mobility solutions, and this part offers valuable insights into how to effectively engage stakeholders and ensure broad societal acceptance. The first chapter by Dominik Schallauer et al. investigates two Austrian SHOW pilot sites, Graz and Pörtschach, by assessing user preferences through a novel "supertester" approach that included experiential elements as well as interviews, questionnaires, and workshops. In their chapter on societal impacts of automated mobility for public transport, Víctor Ferran et al. presents the insights from a modified Delphi study and expert interviews conducted at SHOW pilot sites, assessing direct consequences of AVs on accessibility and equity of public transport, user-perceived safety, job creation and re-skilling as well as indirect effects of automated vehicles such as variation in house prices. In the following chapter, Delphine Grandsart et al. present and discuss three different stakeholders' engagement methodologies that were conducted within SHOW. The Ideathon in Carinthia (Austria), the Hackathon in Thessaloniki (Greece), and the MAMCA (Multi-Actor Multi-Criteria Analysis) workshop in Tampere (Finland) are taken as examples to show that varying stakeholder engagement activities can efficiently generate ideas and validate solutions at a local level enriching the innovation process with novel perspectives. In the last chapter of part III, written by Alexandros Papadopoulos et al., the authors present a comprehensive correlation between automated vehicles' performance and passengers' subjective data from autonomous fleets operation in three SHOW pilot sites (Graz, Madrid, Linköping), each using different technologies and experiencing varying traffic and environmental conditions, confirming a strong correlation between safety and passengers' comfort levels and the vehicles' speed and acceleration profiles.

Finally, *Innovations and Collaborations*, part IV, looks towards the future, showcasing innovative business models and simulations, and highlighting the importance of international collaboration in driving the widespread adoption of automated mobility solutions. The first chapter from Jaâfar Berrada et al. explores several emerging business models for shared automated mobility services through analyzing the experiences and lessons learned from two pilot sites within the SHOW project: Les Mureaux (France) and Monheim am Rhein (Germany). In the second chapter of this part, written by Maria G. Oikonomou et al., the development of an integrated simulation suite that combines elements from the diverse simulations in SHOW, setting the relevant research backbone and enabling their further use by the community, is presented. Lastly, the chapter written by Henriette Cornet et al. underlines the need for international collaboration within research projects for large-scale piloting of shared automated mobility, with lessons learned and recommendations derived from actual collaborations within SHOW between U.S. and Japan.

Through this thematic organization, we aim to provide readers with a holistic understanding of automated mobility, from its technological underpinnings to its broader societal implications. Whether you are a researcher, practitioner, policymaker, or simply someone interested in the future of transportation, we believe that this series offers valuable insights that will inform and inspire.

We would like to extend our sincere gratitude to all the contributors who have generously shared their expertise and experiences in this endeavor.

San Francisco, USA                                                                    Henriette Cornet
Athens, Greece                                                                            Maria Gkemou

# Contents

# Foundations of Automated Mobility

# Novel Vehicle Technologies for Shared Automated Mobility Services with Real-Life Pilots

**Yun-Pang Flötteröd, Marco Münster, Mascha Brost, Sven Ochs, Marc René Zofka, Karl Lambauer, Markus Schratter, and Sebastian Scheibe**

**Abstract** As part of the European project SHOW (GA No 875530), 3 types of innovative automated vehicles, i.e., a U-Shift, a modular vehicle with an interchangeable capsule, a free-moving FZI-shuttle and a VIF-robotaxi, were deployed at 3 sites in Germany and Austria. Two of the vehicles provided both passenger and cargo transport services. The piloting period varied from 7 to 13 months. In total, all automated vehicles travelled 30,000 km and transported 14,000 people. No critical failures occurred. Passengers' enthusiasm and high acceptance of AVs are noteworthy. Further technical challenges still need to be addressed, particularly regarding interaction with other road users, to achieve fully automated operation.

Y.-P. Flötteröd (✉)
Institute of Transportation Systems of the German Aerospace Center, Rutherfordstrasse 2, 12489 Berlin, Germany
e-mail: yun-pang.floetteroed@dlr.de

M. Münster · M. Brost · S. Scheibe
Institute of Vehicle Concepts of the German Aerospace Center, Pfaffenwaldring 38-40, 70569 Stuttgart, Germany
e-mail: marco.muenster@dlr.de

M. Brost
e-mail: mascha.brost@dlr.de

S. Scheibe
e-mail: sebastian.scheibe@dlr.de

S. Ochs · M. R. Zofka
FZI Research Center for Information Technology, Haid-und-Neu-Str. 10–14, 76131 Karlsruhe, Germany
e-mail: ochs@fzi.de

M. R. Zofka
e-mail: zofka@fzi.de

K. Lambauer · M. Schratter
Virtual Vehicle Research GmbH, Inffeldgasse 21A, 8010 Graz, Austria
e-mail: karl.lambauer@v2c2.at

M. Schratter
e-mail: markus.schratter@v2c2.at

H. Cornet and M. Gkemou (eds.), *Shared Mobility Revolution*, Lecture Notes in Mobility, https://doi.org/10.1007/978-3-031-71793-2_1

**Keywords** U-Shift · Automated shuttle · Robotaxi · Shared automated mobility ·
Vehicle technologies

# 1 Introduction

The introduction of automated driving should make driving safer, more comfortable,
and the traffic system more stable. Such technological development can also facilitate the development of more personalized services, and encourage more people
to use shared mobility services, which is one of the keys to achieve a sustainable
transport system. However, the situation in urban traffic is quite complex, and sometimes unexpected. Different weather and seasonal conditions also pose a challenge
to the performances of the sensors and control logics used in automated vehicles
(AVs). For these reasons, numerous demonstrations and pilot operations have been
carried out and many are still running worldwide. In the European project SHOW,[1]
possible AV applications have been demonstrated and evaluated along different use
cases in the shared mobility domain (Show 2024). In the project, the deployed AVs,
offering passenger and/or cargo transport, included automated cars, used as robo-
taxis, automated shuttles (AS) and buses, delivery robots and an automated modular
vehicle.

In this chapter, 3 selected AV types, used at the SHOW test sites are presented after
an overview of the state of art. The pilot activities, the results and the experiences
gained are then discussed. At the end there is a summary and an outlook.

# 2 State of Art

Under a well-defined Operational Design Domain (ODD), AV technology mainly
include perception with sensors, localization, and event-based route/action planning.
These components and control logics assist and guide AVs/AS in recognizing their
surrounding traffic situations and carrying out proper (re-)actions.

## 2.1 Operational Design Domain (ODD)

The ODD for AVs involves the definition of the specific conditions under which the
automated system is designed to operate safely. These conditions typically include
factors such as:

---

[1] https://show-project.eu/.

- Geographical area: the specific region/area where the vehicle is intended to operate. This could range from a city environment to highways or rural areas.
- Environmental conditions: parameters such as weather conditions (e.g. rain, snow, fog), lighting conditions, and road surface conditions (dry, wet, icy).
- Traffic conditions: traffic density level, complexity of road infrastructure, and interactions with other road users e.g. pedestrians, cyclists, and other vehicles.
- Speed range: the speed range at which the vehicle is designed to operate safely, including both maximum and minimum speeds.
- Operational constraints: any specific constraints or limitations on the vehicle operation, such as restricted areas or prohibited manoeuvres.
- Functional capabilities: the capabilities of the automated system, including perception, decision-making, and control, tailored to the identified ODD.
- Regulatory and legal considerations: compliance with applicable regulations and legal requirements governing AV operation within the defined ODD.

Achieving a robust and comprehensive ODD is extremely important to ensure the safe and reliable operation of AVs. Advances in sensor technology, software, and real-world testing are continuously shaping and refining the definition of ODDs to expand the operational capabilities of AVs while ensuring safety and reliability.

## 2.2 Perception and Sensors

AVs use a variety of sensors to detect and interpret their surroundings. These include, e.g., LiDAR, radar, cameras and ultrasonic sensors. LiDAR provides precise 3D mapping of the environment, while radar can detect obstacles, even in poor visibility conditions. Cameras provide high-resolution images that can be used for object detection and tracking. Ultrasonic sensors help to detect obstacles in the vehicles' vicinity. Data from these sensors is processed by powerful algorithms and AI systems to analyse a vehicle's environment and make decisions. However, challenges remain, such as improving sensor range and accuracy and dealing with complex traffic situations. Overall, however, the technology has advanced significantly, enabling AVs to navigate more safely and efficiently. The current AVs' perception and sensor technology do not yet offer the necessary (minimal to zero) failure rate to be operated without a safety driver. This is why there is a safety attendant in AVs in most operations worldwide. A safety attendant must perform various functions, e.g., confirm safe entry and exit, trigger emergency braking in the event of a technical malfunction or trigger driving manoeuvres that cannot be handled by the automation.

## 2.3    Localization

AVs need to determine their positions and adapt their localization requirements based on their automation levels. Certain functionalities only necessitate relative localization, e.g. in relation to lane markings. Others, like some AS in SHOW, require absolute positioning. Absolute positioning entails a coordinate system, e.g., GNSS (Global Navigation Satellite System), providing centimetre-level precision when using a receiver supported by GNSS correction data under ideal conditions. However, problems occur in places where GNSS signals are not accessible, e.g., in buildings, like car parks or bus garages, under bridges, or in tunnels. To resolve temporary GNSS disruptions, sensor fusion can incorporate vehicle data, including speed and acceleration. To mitigate this issue and reduce reliance on GNSS, many AVs use a High-Definition (HD)-map for localization. A HD-map may comprise several layers, with one layer containing the point cloud (Shan et al. 2020) for localization. Vehicle can then localize itself to within one centimetre using the provided point cloud and measured data from the LiDAR sensor (Yurtsever et al. 2020).

## 2.4    Route and Action Planning

The Autoware stack (Autoware Foundation 2024) is an open-source software platform designed for automated driving applications. It provides a comprehensive set of modules and tools to support various aspects of AV development, including perception, planning, and control. Developed by the Autoware Foundation, the Autoware stack encompasses functionalities such as sensor data processing, localization, mapping, path planning, and vehicle control. It aims to offer a flexible and scalable solution that can be customized and integrated into different AV platforms. This software stack is still limited to the own driving lane and offers therefore low flexibility regarding avoiding obstacles on its own lane. The main concept for AS without any steering wheel or brake pedals is the virtual rail, and employs a combination of perception sensors and GPS systems to create a virtual pathway for AS to follow. Due to the fixed trajectory, AS are not capable of avoiding obstacles and the intervention of safety operators is necessary, which disrupts traffic flow and riding comfort.

## 3    Technical Development and Integration

## 3.1    DLR—U-Shift

The DLR has developed the novel vehicle concept U-Shift. It consists of an automated, electric drive module, a driveboard and different transport capsules, which can be flexibly combined with the driveboard. Thus, in combination with different types of capsules, the driveboard enables both people and goods transport and service applications with special capsules. The separation of the driveboard and the transport capsules enables a new kind of flexibility and efficiency for the mobility of tomorrow by choosing an appropriate capsule for each task. The name U-Shift refers to the u-shaped driveboard and the exchangeable capsules (Münster et al. 2021).

Figure 1 gives an illustration of the U-Shift. The driveboard contains most of the technical systems for driverless and electric driving. Cost and resource efficiency are key development goals, which is why expensive systems, e.g. the complex technology for driverless driving, are primarily installed in the driveboard, while the capsules can be built inexpensively. By splitting the vehicle into driveboard and capsules, the driveboard can be highly utilised and operate almost around the clock for maximum efficiency, while the low-cost capsules stay in one place as needed. This allows to reduce costs and resource consumption.

The capsules can be changed quickly, without human operators and "on-the-road", without being tied to external infrastructure, e.g. a crane or similar. The automated change is made possible by a lifting system integrated into the driveboard. With the lifting function, the driveboard moves under the capsule, when it is lowered and lifts it automatically. Secured with a specially developed locking system, the ride can then start. Unlike existing concepts with swap bodies or trailers, the U-Shift capsules offer a very low loading height when parked or when the driveboard is lowered. This allows for fast and easy loading and unloading. The central feature of the automated capsule change without external infrastructure and the low loading height is what makes the U-Shift unique. Special capsules enable a variety of business models and applications—from on-demand or scheduled bus services to parcel delivery services

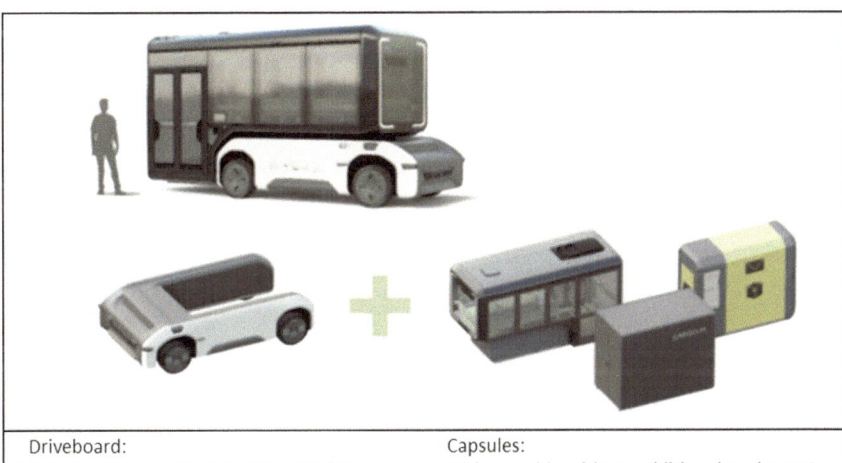

Driveboard:
- Standardised – suitable for all capsules
- Electrically driven and driverless
- Integrated lifting function for capsule change
- Highly utilisable, as driverless (24/7)

Capsules:
- Exchangeable without additional equipment
- Many variants for a wide range of applications
- Demand-oriented application
- "Simple", light and inexpensive

**Fig. 1** U-Shift—driveboard and capsules. *Source* Authors' own pictures

or mobile medical care that can be set up in one place for a required time period. Individual capsules fulfil individual wishes. The scheduling of trips by the driveboard in conjunction with passenger capsules and cargo capsules is conceptually planned by algorithms and independently executed by the system as needed. One of the research demonstrators, U-Shift IV, operates in automated modes where different routes have been learned. The demonstrator was operated in a test vehicle mode at the 2023 Federal Garden Show in Mannheim, Germany (see Sect. 4.1). A detailed safety concept with the points vehicle safety, functional safety, cyber safety was created for the driving license on the test site and tested by the technical supervisory association TÜV Süd.[2]

## 3.2   FZI-Shuttle

The Research Center for Information Technology (FZI) has introduced an innovative concept that deviates from the rigid trajectory limitations of shuttle transport. A major challenge for AS is obstacle avoidance, as they primarily rely on GPS-based routes. Indeed, these AS often have to deviate from their predetermined paths due to various obstructions such as bicycles, cars, garbage bins, or snowdrifts. FZI has adopted a concept, initially developed in the EVA-Shuttle project (Ochs et al. 2023), to address this issue and allow AS to adjust their trajectories dynamically. These required enhancements to various modules, including localization, perception, prediction, manoeuvring, and planning. Not only did the software improvements become necessary, but the hardware upgrades were also made, including the installation of four LiDAR sensors at each corner of the shuttle and the provision of a redundant 360° field of view. This increases the resolution and therefore improves obstacle detection. Moreover, redundancy enhances safety against the failure of individual sensors. Two additional sensors were installed on the rooftop for LiDAR-based localization. A front-facing camera and an onboard unit for ITS-5G communication complemented the sensor setup. The car2x interface has been enhanced with the communication protocols: Signal Phase and Timing, MAP, and Cooperative Awareness Message data. The car body of the EasyMile (EM) shuttle Generation 3 was used.

   The motion planning process starts with the HD-Maps of the target area. For localization, a LiDAR-based localization approach was used since the GPS position in the target area is not sufficient for automated driving functions. The automated driving function facilitates the information provided by the HD-map that needs a centimetre precise global localization. The localization approach uses the SLAM approach, where separate SLAM-map based in LiDAR data was created, and used for the localization during operation. The combination of high-precision position and HD-Map enabled to accurately predict surrounding road users' movements.

---

[2] www.tuvsud.com.

**Fig. 2** The components of the motion pipeline of the FZI, that are all handled by the FZI-shuttle itself. During the public operation all components were continuously updated. *Source* Authors' own pictures

In Fig. 2, the cyan depiction illustrates the HD-map based on the lanelet format (Ochs et al. 2023). The lanelet format represents foremost the road boundaries, but also semantic information like right-of-way. The extracted lanelets serve as the basis for the driving area marked in yellow. This driving area, subject to dynamic adjustments based on manoeuvre decisions, provides a flexible space. To safely and efficiently navigate through intricate scenarios with dynamic obstacles, a strategy that breaks down the complexity into more manageable decision-making units, the so-called "Traits". These "Traits" represent specific considerations, e.g. narrowing the driving area in response to oncoming traffic or expanding the driving area when confronting with a parked vehicle in the same lane.

To reconcile conflicting directives, posed by these Traits, a resolver mechanism has been introduced. When constructing the planner's environmental model, a further distinction is made between dynamic and static obstacles. Static obstacles are represented in an occupancy grid, the so-called costmap. Within this costmap, polygons are derived to encode static obstacles. Meanwhile, dynamic obstacles are directly treated as polygons during perception pre-processing and then transmitted to the planner in this polygonal form. The used motion planning pipeline is based on the Particle Swarm Optimization (PSO) and developed by FZI. This approach also stands out for being easily adaptable to new scenarios. The parallel calculation enables fast planning cycles. Following the PSO principles, the trajectory planer first generates a swarm of initial trajectories that are optimized afterwards. The output trajectory is determined by the particle with the best performance within the optimized swarm, as shown in the planning step in Fig. 2.

The motion planning pipeline is integral to an on-demand service. The ioki booking system (Ioki 2024) facilitated this at the AS, providing a smartphone app and backend for safety operators. With dynamic rerouting, FZI-shuttles were able to respond to new bookings mid-route, expanding their capabilities and enabling ride pooling in the target area. This approach has been used in the FZI-shuttles, already driven more than 3,000 km safely and entirely autonomously in suburban daily traffic.

### 3.3  VIF-Robotaxi

Virtual Vehicle (VIF)'s robotaxi is a standard commercial passenger car (Ford Fusion) equipped with various sensors to detect and predict what is happening around the vehicle for safely navigating through different situations. VIF enhanced the capabilities of the AV in SHOW to safely navigate complex bus terminals, including manoeuvring through crowded areas with pedestrians, cyclists, buses and trams. Figure 3 shows all installed LiDAR, radar, camera and GNSS sensors and the interface to the actuators (lateral and longitudinal control) of the vehicle, the various computing platforms and the location of each sensor in the right image. Different types of sensors are used to increase the operating robustness of the system.

In the vehicle, Autoware has been used as the basis for the driving functionality (Kato et al. 2015). The open-source automated driving software stack is an ideal platform for research vehicles since it contains a variety of basic functionalities and can be used in various projects without license restrictions. However, numerous modifications and integration efforts have been required to use the software stack in the vehicle. Figure 4 shows the vehicular system architecture and the software components used. An industrial PC with Ubuntu 22.04 and ROS as middleware is used. The necessary drivers for the sensors/actuators and the applications communicate via ROS. The applications contain the modules described in the state of the art.

To plan a path from points A to B, the HD-map of a road network is required (Poggenhans et al. 2018). In addition to path planning, the HD-map is used to predict the trajectories of other road users for the obstacle avoidance algorithm, based on the road geometry. Figure 5 shows the layer of the point cloud, used at the Graz pilot site, and the high reflection of reality. In particular, Autoware uses the Lanelet2 map format. The map format contains information about road geometry, driving directions, traffic light positions, speed limits, etc. This data, together with the predictions

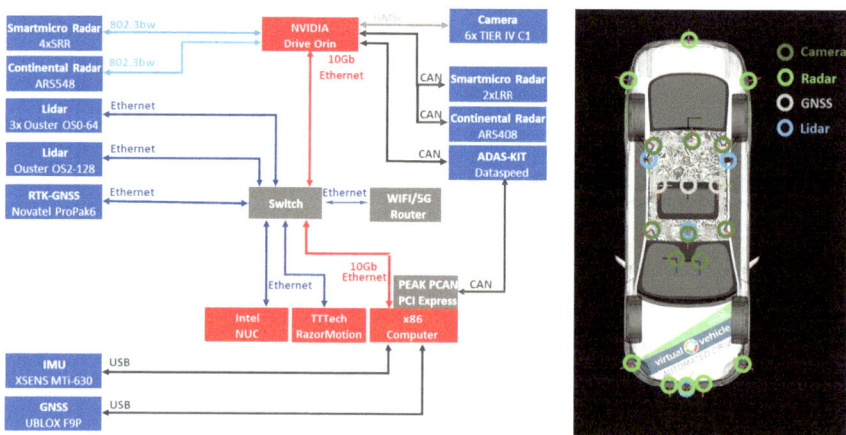

**Fig. 3**  The used components in the VIF-robotaxi demonstrator. *Source* Authors' own pictures

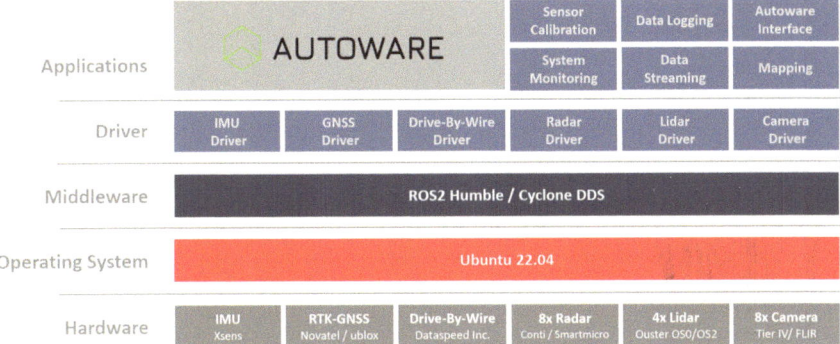

**Fig. 4** System architecture of the VIF-robotaxi demonstrator. *Source* Authors' own pictures

of other road users, is used to generate the trajectory for the automated driving. In addition to the routing functionality for normal vehicle driving function, adaptations were required for the passenger service. VIF adapted the software to store fixed stop positions in the HD-map, which can be selected via a web service to send commands to the vehicle.

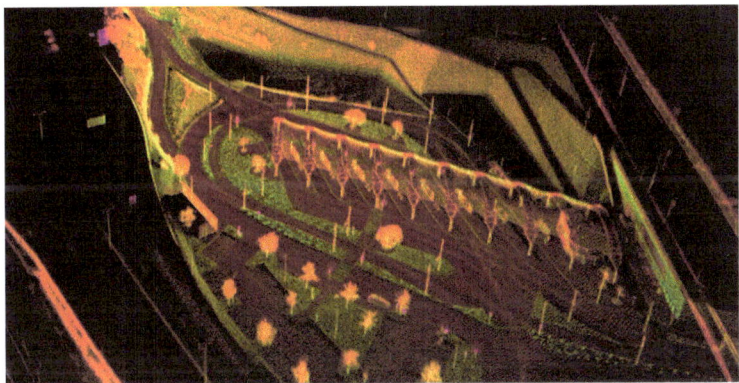

**Fig. 5** Point cloud used for the LiDAR-based localization for the bus terminal at the Graz pilot site, highly reflecting the actual situation on site. *Source* Authors' own pictures

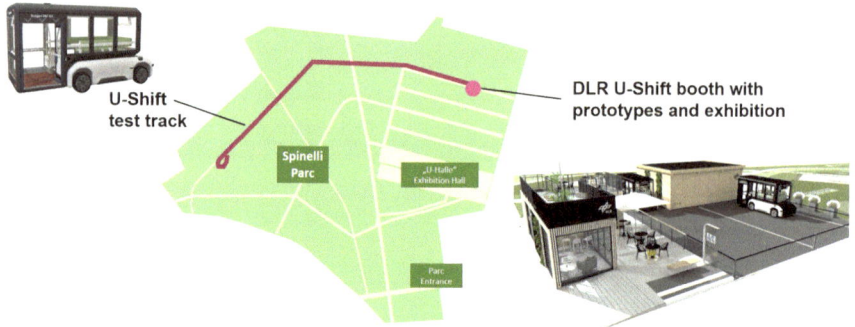

**Fig. 6** U-Shift test track on the BuGa 23. *Source* Authors' own pictures

## 4 Demonstrations and Evaluation Results

### 4.1 DLR—U-Shift

The U-Shift concept was tested via several DLR prototypes at the Federal Garden Show in Mannheim (BuGa 23[3]) with 85,000 visitors. The garden show area was not a public street space, but a demarcated area that was nevertheless characterised by a lot of public foot traffic and to a small extent also other types of traffic. In the period from April to October 2023, the vehicle drove on a test track and it was presented to visitors as part of an accompanying exhibition. The prototypes were two driveboards of different functionalities and development stages, two passenger capsules, one cargo capsule and one multi-use platform. The test route was an approximately 2 km long asphalt road (see Fig. 6) and was not exclusively for the U-Shift vehicle, but was shared with pedestrians and small electric road trains that regularly ran on the same route. The automation functions were constantly developed during the test period. Passengers initially experienced a driver, controlling the vehicle via joystick and drive and steer system. After comprehensive test drives in automated mode, passengers could experience the U-Shift IV in automated driving. The U-Shift IV has LiDAR, camera, radar, ultrasonic and GPS sensors. Different approaches were taken for automation. The main demonstration application was an automation approach, in which path planning on a fixed route was specified with the use of real-time kinematics (RTK) GPS and supplemented by a LiDAR-based collision warning system. A LiDAR-based SLAM was investigated as well. In total, no critical issues raised and more than 10,000 visitors took a ride on the U-Shift.

---

[3] "BuGa" is short for "Bundesgartenschau", the German national garden show https://www.buga23.de/englisch/.

## 4.2 FZI-Shuttle

The concept of free trajectories was a fundamental element of the testing methodology applied at the Karlsruhe test site during the period from December 2022 to December 2023. The site consisted of a controlled area with restrictions and a larger suburban zone. The controlled area comprised a 2 km stretch of road and a 5,000 m² open space, specifically designed for experimental tests. In contrast, the primary testing area was located in a suburban in the south of Karlsruhe, Germany. Within this suburban environment, the FZI-shuttles travelled on public roads and had interactions with a variety of road users, including conventional vehicles, cyclists, e-scooter riders, and pedestrians.

In the depicted scenario, a cleaning truck temporarily blocked the designated lane of the FZI-shuttle, forcing the shuttle to adapt to the situation by seeking an alternative route. During this process, the shuttle encountered oncoming traffic, which temporarily needed a pause in its manoeuvres. However, it is noteworthy that the FZI-shuttle automatically and efficiently navigated around the obstructing vehicle without any manual intervention from the safety operator, exemplifying the concept of "free trajectories" in action. In this area, the FZI-shuttles made over 1,000 public trips, a total of over 3,000 km. The maximum driving speed was 20 km/h according to the restrictions imposed by the granted permission for operation. The FZI-shuttles were required to adhere to a newly developed safety concept. This concept stipulates a maximum speed that is based on the distance to other obstacles. Based on this concept, Fig. 7 shows the exemplary speed profile of a trip, where red and yellow colours represent a low velocity. This is due to the fact, that these streets are narrow and many vehicles were parked on the roadside. As soon as the FZI-shuttles left the narrow streets, the colours changed to blue/green, which means that the maximum travelling speed of 20 km/h was almost reached.

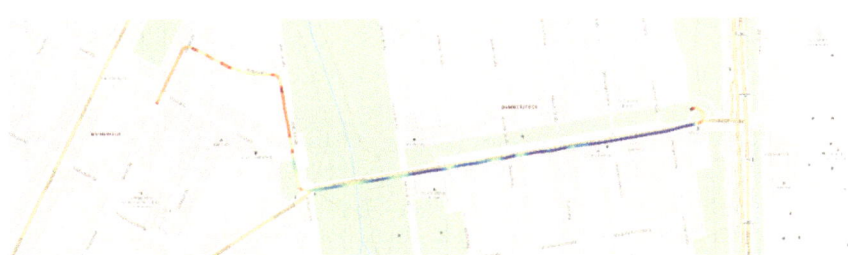

**Fig. 7** The speed profile of one trip with the FZI-shuttles in the target area, where red/yellow and blue/green represent a low and a higher velocity respectively. *Source* Authors' own pictures based on OpenStreetMap (OSM 2024)

### 4.3 VIF-Robotaxi

The piloting was carried out at the Graz pilot site with the focus on connecting peri-urban regions to intermodal mobility hubs in mixed traffic. The pilot was located on the southern outskirts of the city, about 4 km from the city centre. In this area, there are a shopping centre and the public transport (PT) hub "Puntigam", around 1 km apart and connected via the VIF-robotaxis. The PT hub "Puntigam" is served by 6 public bus lines and 1 tram line, and is also connected to the nearby urban railway stations. In this urban scenario, the robotaxis stopped at the bus terminal, picked up people upon request, drove through the public stops, where there were many pedestrians, and then arrived the shopping centre. With the help from the traffic infrastructure (e.g., guiding through traffic lights), vehicles could perform actions in an automated way (at slow speed).

Two robotaxis were in use at this site: one Ford Fusion and one Kia e-Soul. The Kia Soul from AVL was updated to support the same capabilities (automated driving software) as the Ford Fusion, and the connectivity with the SHOW Data Management Portal was established in real time via MQTT for both vehicles. The deployment of the infrastructure components (AwareAI camera, C-ITS RSU) was also carried out, configured and tested (Yunex). The camera detects available bus bays for the AVs' passage. Besides the 4G/5G network, V2X communication based on ITS G5 was used, setup in close cooperation with KAPSCH TrafficCom and Yunex.

The public pilot phase was from October 2022 to September 2023. Around 520 passengers were transported in total. The passenger rides were conducted during public weeks and other events open to all visitors.

### 4.4 Results and Lessons Learned

All AVs were examined, inspected and approved with their technical functions and safety concepts by the responsible local authorities according to the respective regulations, before operating at the pilot sites. Due to the different site characteristics, weather conditions and regulations, the AVs' ODDs were different, but the main factors, as mentioned in Sect. 2.1, were all considered. Safety operators were required at all sites during operation. No critical technical failures occurred, as expected. Table 1 provides the overview and comparison of the three AVs.

During the 180 days of test operation of the U-Shift at the Federal Garden Show 2023, around 2,800 km were travelled and 10,000 people were transported. After initial adjustments to the vehicle control system and basic functions, as well as fine-tuning the controls, the U-Shift's operation has become increasingly robust. Operation on the road, which was shared with pedestrians, led to various conflicts that were accident-free. Scenarios arose in which people came so close to the vehicle that it was not possible to continue driving and the danger zone was cleared again by means of an acoustic warning signal. Unfortunately, there were also situations in

**Table 1** Overview of the characteristics, operation conditions and performances of the AVs

| Item | FZI-shuttle | DLR—U-Shift | VIF-robotaxi |
|---|---|---|---|
| Vehicle type | 2 modified EM-shuttles | 1 module vehicle | Ford Fusion, Kia e-Soul |
| Sensor types | LiDAR, camera, radar | LiDAR, camera, radar, ultrasound | LiDAR, camera |
| Capacity (people) | 6 | 8 (up to 15) | 5 |
| Localization method | LiDAR, HD-maps | RTK, GPS, LiDAR | LiDAR, HD-map (point-cloud) |
| Path/routing logic | Moving freely | Fixed route with virtual rail | Fixed route |
| Approved operation area | Resident area; KIT campus | Exhibition area; KIT campus | Peri-urban area |
| Route length | 4.5 km | 2 km | 1 km |
| Maximum speed | 20 km/h | 15 km/h | 56.7 km/h |
| Operation type | DRT | Fixed schedule | DRT |
| Stop types | Virtual stops | Fixed stops | Fixed stops |
| General restrictions (all) | Only in the target areas; daytime only; no heavy rain/snow/ice | | |
| Operation period | 2022.12–2023.12 | 2023.04–2023.10 | 2022.10–2023.09 |
| Transport types | People, parcels | People, parcels | People |
| Kilometres travelled | 338/2,662 (operation/ tests) | 2,800 | 900/23,630 (operation/tests) |
| Passengers transported | 3,477 | 10,000 | 520 |
| Hard breaking event | 140 | 5 (purposely incited by people) | 3 |
| Conflicts–near miss incidents | 100 | 20 (closely to pedestrians) | 0 |
| Illegal overtakes | 69 | 0 | 0 |
| Road accidents | 1 | 0 | 0 |

which people deliberately ran in front of the vehicle, which led to emergency braking. The important finding is the enthusiasm of the passengers and the high acceptance of automated vehicles.

During the operation of the FZI shuttle, two main findings emerged. Firstly, both safety drivers and passengers felt secure and found the ride very pleasant. This is partly due to the automatic obstacle avoidance and therefore lack of interventions by safety operator. Secondly, we were able to increase traffic throughput with the advanced driving function. However, there were recurring issues with vehicles and cyclists merging in front of the shuttle or disregarding right-of-way rules, which led to a higher number of critical situations compared to the U-shift and VIF-Robotaxis. This can be due to the low maximum speed of the FZI shuttles.

During almost 400 test drives with the VIF-robotaxis, there were few non-critical individual situations, in which the safety driver intervened in the operation. One occurred when a vehicle occupied the robotaxi's stop. Although the robotaxi recognized the unexpected object and stopped behind it, the robotaxi blocked the bus lane and impaired bus operations. Another one happened when the robotaxi confronted with a situation, in which a bus travelled against a one-way bus lane for reaching a parking lot. The bus drove towards the robotaxi since no one gave the bus any instructions. These cases point out that AVs need to be able to perform more flexible manoeuvres, e.g. driving in the opposite direction away from another object, instead of just stopping or deviating from the predetermined lane/path, since unexpected situations can occur anytime in a complex traffic environment.

The above issues imply that there is room for improvement to draw other road users' attention to moving AVs, and to enhance reaction alternatives and decision-making of AVs. The capability to better handle more critical weather and pavement conditions, e.g. heavy rain and snow and icy roads, should also be further addressed.

## 5 Summary and Outlook

Overall, the U-Shift, FZI-shuttles and VIF-robotaxis performed well during the demonstration and evaluation periods, having presented several novel approaches and their practicability for future shared automated mobility. No critical issues have arisen. Challenges still persist, particularly when interacting with other road users and at intersections that are difficult to see. In order to start fully automated operation, many small technical hurdles still need to be overcome over the next few years and the flexibility in action taking needs to be further enhanced to make sophisticated decisions for driving safety and efficiency.

Currently, there are no further AV/AS operational plans and projects at the Karlsruhe and the Graz test site. But the FZI continues to develop automated driving functions with the FZI-shuttles focusing on teleoperation and remote fleet-management. VIF is also continuously developing and enhancing automated driving functions to deal with various traffic situations. Further work on various projects in the U-Shift project landscape is planned. As part of the U-Shift technology transfer project, the aim is to obtain approval for the use of the U-Shift IV on public roads. Besides, various vehicle components will be further developed and the technology transfer to small and medium-sized enterprises (SMEs), medium-sized companies and large companies will be continued.

**Acknowledgements** The work presented in this manuscript has been funded by the SHOW project which has been made possible by funding from the European Union's Horizon 2020 Research and Innovation Programme Under Grant Agreement no. 875530.

# References

Autoware Foundation (2024) Autoware [Online]. Available: https://github.com/autowarefoundation/autoware. Accessed 15 March 2024

Ioki (2024) IOKI Platform [Online]. Available: https://ioki.com/en/platform/. Accessed 6 March 2024

Kato S, Takeuchi E, Ishiguro Y, Ninomiya Y, Takeda K, Hamada T (2015) An open approach to autonomous vehicles. IEEE Micro 35(6):60–68. https://doi.org/10.1109/MM.2015.133

Münster M, Brost M, Hahn R et al (2021) U-Shift vehicle concept: modular on the road. In: Proceedings of 21 internationales stuttgarter symposium, pp 333–346

Ochs S, Grimm D, Doll J-D et al (2023) Stepping ahead with electrified, connected and automated shuttles in the test area autonomous driving BW, TechRxiv. https://doi.org/10.36227/techrxiv.170327243.35964930/v1

OSM (2024) OpenStreetMap [Online]. Available: https://www.openstreetmap.org/. Accessed 7 May 2024

Poggenhans F et al (2018) Lanelet2: a high-definition map framework for the future of automated driving. In: Proceedings of 21st international conference on intelligent transportation systems (ITSC), pp 1672–1679

Shan T, Englot B, Meyers D, Wang W, Ratti C, Rus D (2020) LIO-SAM: tightly-coupled lidar inertial odometry via smoothing and mapping. In: Proceedings of IEEE/RSJ international conference on intelligent robots and systems (IROS), pp 5135–5142. https://doi.org/10.1109/IROS45743.2020.9341176

Show (2024) SHared automation Operating models for Worldwide adoption: mega sites–Germany, Grant Agreement no. 875530, 2024 [Online]. Available: https://show-project.eu/mega-sites-germany/. Accessed 05 March 2024

Yurtsever E, Lambert J, Carballo A, Takeda K (2020) A survey of autonomous driving: common practices and emerging technologies. IEEE Access 8:58443–58469. https://doi.org/10.1109/ACCESS.2020.2983149

# Remote Supervision Strategies for Automated Vehicles Fleets: Three Real-Life Operational Case Studies

Henriette Cornet, Sofia Pavlakis, William Levassor, and Nicolas Morael

**Abstract** Public Transport Operators (PTOs) are crucial actors in developing and deploying shared automated mobility services with their role of ensuring service quality, passenger safety, and cost-effectiveness. This chapter underscores the pivotal role of remote supervision in facilitating the seamless operation of automated vehicle fleets from the point of view of PTOs. Drawing from real-life deployments of the EU-project SHOW (GA No 875530) and other initiatives, it highlights the necessity of learning from everyday situations to enhance operational efficiency and safety. Additionally, the vision for scalable supervision centres tailored to higher levels of automation is outlined, along with the imperative for standardization at the regional level to be able to scale the integration of automated vehicles into future public transport systems.

**Keywords** Automated mobility · Fleet supervision · Public transport operators · Road safety · Public transport services

H. Cornet (✉)
University of San Francisco, 2130 Fulton St., San Francisco, CA 94117, USA
e-mail: hcornet@usfca.edu

S. Pavlakis
Rms GmbH, Am Hauptbahnhof 6, 60329 Frankfurt Am Main, Germany
e-mail: sofia.pavlakis@rms-consult.de

W. Levassor
Beti, 30 Avenue Gambetta, 26260 Saint-Donat-sur-l'Herbasse, France
e-mail: william.levassor@beti.team

N. Morael
Transdev Group, 3 Allée de Grenelle, 92130 Issy-Les-Moulineaux, France
e-mail: nicolas.morael@transdev.com

# 1  Introduction

Public Transport Operators (PTOs) are at the forefront in the development and deployment of shared automated mobility services. With a primary responsibility to ensure the quality of service, passenger safety, and cost-effectiveness, operators play a crucial role in shaping the future of transportation with automated vehicles (AVs). In this context, the role of remote supervision emerges as a pivotal component in ensuring the seamless operation of automated vehicles fleets. By transitioning from traditional driver-centric models to centralized supervision, operators can potentially alleviate the multifaceted burden on drivers, including the significant portion of fleet costs attributed to their wages and the inherent unreliability such as sickness or constraints with driving schedules. This shift enables operators to maintain oversight of multiple vehicles from a single control centre, fostering greater efficiency and flexibility in operations. Embedded in the SHOW project,[1] this chapter (i) presents the learnings from a supervision centre of AVs without safety driver on board, (ii) outlines a possible vision for the role of PTOs in supervision with higher levels of automation, and (iii) underscores the imperative for standardization at the regional level to realize the full potential of remote supervision in advancing shared automated mobility services.

Section 2 summarizes the most recent investigations towards remote supervision of AV fleets, while Sect. 3 focuses on three real-life case studies from PTOs active in SHOW. Section 4 discusses the learnings and next steps.

# 2  Background and Context

Ensuring effective public transport provision demands significant organizational endeavours, planning, financial support from the public, and the seamless coordination of millions of passengers and staff members within extensive systems. In this context, and independently from AVs, PTOs have the responsibility to ensure the safety of the passengers and other road users, guarantee the quality of service (following the strict specifications set by Public Transport Authorities (PTA) who mandates them), and ensure cost-effectiveness since efficient resource allocation stands as a pivotal factor in the daily operations of such systems (Hörcher and Tirachini 2021).

Taking into account the three primary responsibilities of PTOs regarding safety, quality of service, and cost-effectiveness, the integration of automated vehicles (AVs) presents a significant opportunity to enhance performance across all these domains. Firstly, AVs contribute to heightened safety standards by leveraging automated driving functions, thereby mitigating risks associated with driver fatigue and distractions through the utilization of advanced sensor technologies. Secondly, AVs have the potential to enhance passenger satisfaction by enabling increased service

---

[1] https://www.show-project.eu/.

frequency, improved reliability, and broader area coverage, particularly in regions where deploying a traditional fleet proves cost-prohibitive due to driver wages and driver shortages. Lastly, the adoption of AV technology holds the promise of long-term cost reduction, as labour costs associated with drivers constitute a substantial portion, approximately 40% of fleet operation expenses (Ongel et al. 2019). Therefore, by embracing AVs, PTOs can simultaneously elevate safety, service quality, and cost-effectiveness, fostering a more efficient and sustainable public transportation ecosystem.

However, the belief that deploying fleets of AVs imply no human involvement due to the absence of a human physically present in the vehicle is a misconception (Cooke 2006). Personnel will be required to supervise remotely AVs fleets and by doing so 'keeping a human in the loop' (Amador et al. 2022). The SAE J3016™ standard (SAE International 2021) defines *remote supervision* as the act of "monitoring and responding as needed to perform the Dynamic Driving Task (DDT)" through personnel remotely located, i.e., not inside the vehicle. The standard also defines the term *remote driving*, where remote personnel "has the authority to overrule the ADS (Autonomous Driving System) for purposes of lateral and longitudinal vehicle motion control". This chapter focuses on remote supervision, taking the position that remote driving is not desirable at the moment, considering the risks associated with shared liability between the PTO and the ADS provider, and the confusion that arises when roles are mixed, potentially allowing a PTO to assume the responsibilities of an ADS.

A fleet supervision centre is similar to those existing for non-automated vehicle fleets, for instance for monitoring bus fleets in a city. This centre comprises essential hardware components such as computers, communication equipment, surveillance systems, GPS trackers, and emergency communication systems, facilitating real-time monitoring, communication, and coordination of vehicle operations. Trained supervisors, technicians, dispatchers, and an emergency response team comprise the personnel. However, managing AV fleets presents new challenges for fleet supervision, including ensuring real-time situation awareness (addressing latency issues to ensure timely data transmission), preventing information overload and fatigue among remote supervisors, and addressing scalability and standardization concerns to enable replication across different locations. While other challenges such as remote *driving* and Human–Machine Interface (HMI) are pertinent, this chapter focuses on remote *supervision* with its technical and workforce requirements, vision for scalability, and the need for regional standardization.

# 3    Remote Supervision Strategies and Visions from SHOW Pilot Sites

The following subsections describe learnings and recommendations from three PTOs and strategic partners involved in SHOW at various locations: (i) Transdev (www.transdev.com) PTO in Les Mureaux, France; (ii) beti (www.navette-autonome.fr) PTO in Crest, France; and (iii) rms (www.rms-consult.de) strategic consultant working closely with the regional PTAs and local PTOs in the Rhine-Main-Region (Germany).

## 3.1    Building a Supervision Centre for First AVs Operations Without On-Board Safety Driver in Europe

The transition from trials to commercial services marks a pivotal moment for Automated Mobility, particularly when the on-board operator is removed, signifying a crucial step towards fully autonomous operations. Demonstrating the viability of the Automated Vehicle (AV) business case axes on the effective supervision of a fleet of shuttles by a single supervisor. To achieve this, continuous testing and improvement of technologies are essential in preparing for the deployment of AV commercial services. Key areas of focus include remote supervision, which serves as a fundamental enabler for AV commercial services, ensuring safety and service efficiency and ultimately, advancing towards fully driverless operations to develop operational expertise.

Until 2020, for AVs to be tested on public roads, legislation universally required that a safety driver must be inside the vehicle, ready to intervene if a disengagement was requested (Mutzenich et al. 2021). Since then, changes to many European, U.S. states and UK regulations have enabled a remote operator to assume this role.

At the SHOW pilot site in Les Mureaux, France, located on the private grounds of aerospace company Ariane Group, Transdev became the first operator in Europe to operate three EasyMile shuttles (EZ10 Gen3) without a safety driver on board. Despite the absence of on-board safety personnel, the service, system, and passengers' experiences are supervised by an OCC (Operating Control Centre) on site with remote supervision.

The goal of the pilot site is to test the feasibility of supervising remotely a fleet of three automated shuttles with a single supervisor from the OCC, while ensuring the safety and performance of the service.

The remote supervisor oversees all alerts from the system, including those from the vehicle platform, the ADS system, and potentially connected infrastructure. Alerts are prioritized based on the actions required by the remote supervisor. For instance, he/she can request the vehicle to stop at the next safe zone and/or authorize specific manoeuvres. When such actions are necessary, a video feed is automatically displayed on the supervision screen. It is important to note that the safety of driving functions is always managed by the ADS system within the vehicle. This means that the remote supervisor cannot perform an emergency stop remotely but can only request the vehicle to stop at the next safe zone. The ADS system receives the request and stops the vehicle where it is safe. All of these functionalities aim to ensure passenger safety and service quality while maintaining clear delineation of liability between the ADS and the remote supervisor. In addition to the management of the service and the system, the remote supervisor is managing the passengers' experience. He/she can inform passengers through on-board screens or directly communicate with the passengers with a microphone. He/she can also manage comfort inside the shuttle by e.g. adjusting the lightening and the temperature.

When the ADS and the supervision system cannot manage specific situations, remote supervisors can send field agents to intervene on the vehicle and/or to assist passengers.

In Les Mureaux, remote operators undergo training provided by Transdev and EasyMile, initially as on-board safety operators to understand the technology and how the vehicle behaves in real conditions. In the next phase, after several months of operations, they receive training as remote operators. During this transitional period, on-board operators are still required. After a few weeks, once training is fully completed and validated, the driverless service is launched.

The automated shuttle is equipped with four external cameras, along with one camera inside the vehicle (Fig. 1). In the event of any issues, the operator has access to video streams from the 360° cameras placed on the vehicle (Fig. 2) and has the capability to communicate directly with passengers to address concerns promptly.

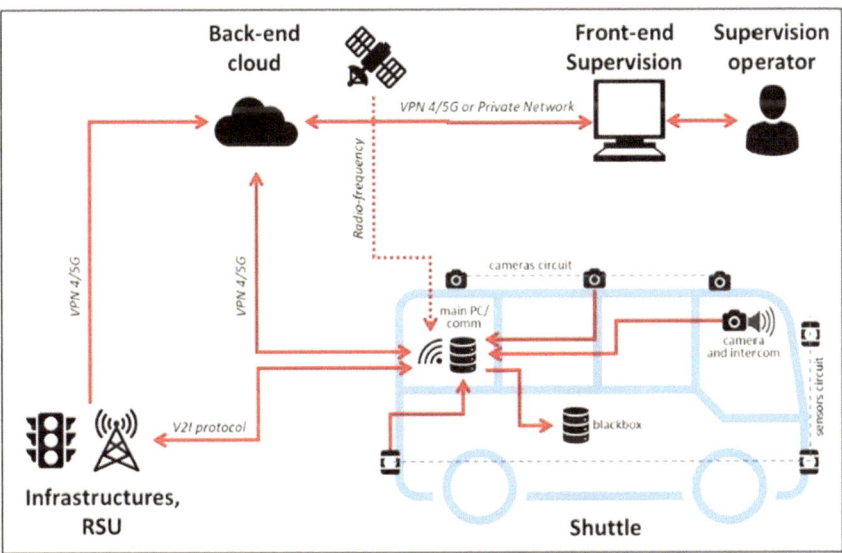

**Fig. 1** Simplified system architecture for remote supervision at Les Mureaux. *Source* Transdev

**Fig. 2** Operating Control Centre (OCC) at Les Mureaux. *Source* Transdev

Removing the on-board operator brought many new challenges to continuously ensure safety, quality of service and passengers' experience. Following lessons learned have been gathered from the pilot Les Mureaux so far.

- Roadside Assistance response time

While the technology will continue to mature over time, there will always be a need for roadside assistance teams to intervene when necessary. In Les Mureaux, the focus was on efficiently organizing these teams to quickly resolve issues. This may involve deploying another vehicle from the service to ensure passengers reach their destination within an acceptable timeframe. The response time of roadside assistance teams will be crucial in managing the quality of service for operations at scale.

- Operational Safety

To ensure operational safety, Transdev implemented comprehensive risk mitigation strategies and stringent safety protocols to enhance safety measures and improve operational reliability. For example, analysing the driving behaviour of other road users in critical areas such as roundabouts required progressively adapting the AV shuttle's speed and/or route to maximize safety and traffic flow.

- Supervision camera system with latency monitoring

An important lesson learned is the necessity for the supervision camera system to be fully integrated with the monitoring system, including a latency monitoring system. This ensures that the supervision operator always sees what is happening in real-time on the road. This is crucial not only for taking immediate actions but also for managing passengers' journey experiences (boarding, disembarking, communication, etc.).

- Stable communication system

Another significant insight is the importance of a stable communication system between the OCC and passengers, complemented by indoor video flow. This becomes especially crucial before widespread adoption of AV technology to insure passenger trust.

- Business Case viability

One of the primary benefits of autonomous mobility technologies is the reduction in operational costs. In Les Mureaux, Transdev aimed to demonstrate the viability of this business case by assigning one supervision operator to oversee three shuttles. As the number of vehicles supervised by one remote operator increases, so does cost efficiency. Scaling up, the Total Cost of Ownership (TCO) of an automated mobility service should be lower than that of a human-driven service. This will enable the extension of operational areas (such as serving low-density, peri-urban, and rural areas) and service hours throughout the day, including services during the night.

The next steps of the pilot site are to extend the service to a broader urban context, testing for instance last-mile connections on open roads.

## 3.2 Future Functionalities of Remote Supervision: The Concept of an Hypervision Platform

The Crest pilot site represents a significant effort to mitigate transportation gaps by offering residents a complimentary alternative to private cars. With a safety driver accompanying each journey, passengers can embark on-the-spot without the need for prior booking. This pilot site, with a fleet of four NAVYA (since 2023, GAMA) electric shuttles, aligns closely with the broader vision of connected and decarbonized mobility, aimed at reducing both private car usage and transport-related $CO_2$ emissions.

The Crest pilot site focuses on addressing transportation challenges in rural areas, where residents typically exhibit a high dependency on private cars or face mobility constraints, especially the elderly population. Despite the cost-intensive nature of deploying mobility services in such areas with low demand, it remains imperative for territories and local operators to ensure inclusivity and accessibility to transportation services, thereby facilitating access to jobs and other societal duties.

Automated mobility presents a promising solution to address challenges such as driver shortages, extended service hours, and safety enhancements. However, it is essential to recognize that the transition to AVs does not eliminate the need for human oversight. The vehicles will drive and avoid collision, but the human element remains crucial in managing the system and dealing with operators' commitments: ensuring passenger security and proposing the best quality of service at any time, particularly in addressing hazardous events that may not be fully addressed by the self-driving technology embedded within AVs.

Indeed, from an operational perspective, drivers play a multifaceted role that extends beyond driving. Their responsibilities include ensuring passenger security, providing information and comfort, maintaining service quality, and managing

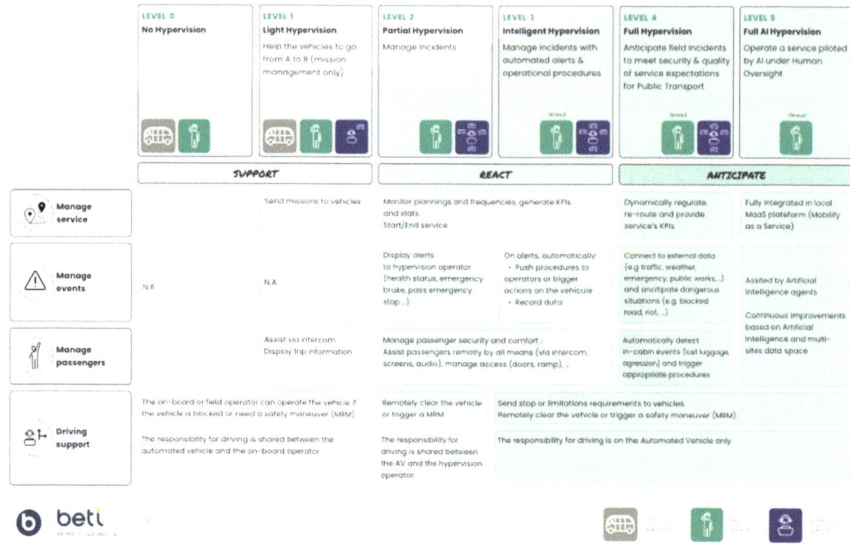

**Fig. 3** The five levels of hypervision for automated mobility fleets. *Source* Beti

unforeseen or out-of-sight events. These events may involve incidents such as passengers attempting to exit between two stops while the vehicle is stopped due to an incident, passenger falls, or operational situations that could jeopardize passenger safety if the vehicle proceeds (such as riots, floods, or fires along the intended route). These scenarios underscore the need for comprehensive operational security measures that surpass the capabilities of driving technology alone.

Currently, only a fraction of these roles and capabilities is visible in AV pilots. So far, the remote supervisor has the ability to send and monitor missions, communicate with passengers remotely, and validate autonomous driving path choices from a distance. These functionalities will gradually develop to achieve commercial-grade quality of service.

In this context, beti, the operator in Crest, has developed a vision for a 'hypervision platform' to facilitate the daily operation of the automated mobility services. Drawing inspiration from various mobility services—from elevators to driven fleets of buses and aeronautic—this platform incorporates functionalities aimed at monitoring and optimizing service performance, anticipating and mitigating downtimes and safety issues, enhancing the user experience, and supporting the driving system to reduce emergency stops or minimum risk manoeuvres (MRM).

The required functionalities of the hypervision can be classified in 5 different levels, making an echo and complement to the SAE levels of driving automation[2] (Fig. 3).

---

[2] With a taxonomy for six levels of driving automation, Society of Automobile Engineers (SAE) J3016 defines the SAE Levels from Level 0 (no driving automation) to Level 5 (full driving automation).

Currently, the remote supervision capabilities of most AV pilots are located at a level of support and reaction (for level 0 to level 3) of hypervision. At term, it is expected that full AI hypervision will be reached, enabling an AV service to be pilot by an AI under human oversight.

On the long term (level 5 of Fig. 3), the hypervision platform aims to enhance the following groups of functionalities with advanced services:

1. Manage service:

Often referred as fleet management or Computer-Aided-Dispatch, the service will be fully integrated in the local MaaS (Mobility as a Service) platform contributing to a seamless mobility experience for the passengers.

2. Manage events:

Sometimes referred as technical supervision, the platform will autonomously monitor system health, generate alerts, and connect with local information and emergency services (e.g., blocked road, riot).

3. Manage passengers:

The platform will assist in informing and support passengers, managing access and comfort, and detecting in-cabin events to reduce reaction time and enhance the overall user experience.

4. Driving support:

Sometimes referred as tele-assistance or tele-operation. Through functionalities such as sending requirements to limit vehicle speed and assisting in path planning, the platform will support the driving system to minimize emergency stops or MRM and downtimes for instance when the vehicle is waiting for a validation.

In summary, beti's hypervision platform concept aims at advancing the operational capabilities of automated mobility services, providing operators with the tools necessary to ensure safety, efficiency, and passenger satisfaction for commercial services.

## 3.3   The Need for Standardization at Regional Level

The Rhein-Main-Verkehrsverbund (RMV)—as the regional public transport authority (PTA)—considers automated shuttles to be an important component for meeting the agreement of mobility and climate protection in future public transport (PT) (Rhein-Main-Verkehrsverbund 2022).

The Rhine-Main Region consists of several large cities, such as Frankfurt-am-Main or Wiesbaden (>200,000 inhabitants), and a high number of suburban cities and numerous rural areas—altogether with a total population exceeding 5.8 million inhabitants. These rural areas in the region often have a lack of availability of PT.

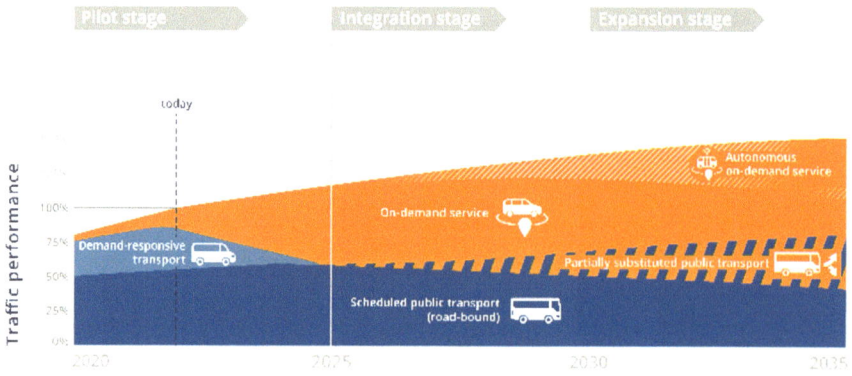

**Fig. 4** Vision for autonomous on-demand services in the Rhine-Main region. *Source* Rms

There, automated vehicles can be efficiently used as on-demand services and run upon request.

A digital driver-based demand-responsive transport (DRT) service was available prior to SHOW not only in Frankfurt but also in 9 other cities and municipalities within the authority of the RMV. However, this service is cost-intensive for the PTA since it requires a large number of vehicles and therefore a high number of cost-intensive drivers. For this reason, rms (a consulting company working with PTOs) was mandated to supervise the replacement of some of the vehicles by automated shuttles in the SHOW pilot site of Frankfurt where two EasyMile shuttles ran an automated on-demand service that could be booked through a Smartphone App. More information about the operational service can be read in Rhein-Main-Verkehrsverbund Servicegesellschaft mbH (2024).

Based on the success of SHOW and other projects conducted in the region, RMV has the vision to expand the use of on-demand services, which a part is to be provided by automated shuttles (Fig. 4).

At the Frankfurt pilot site and within the framework of other projects besides SHOW, a supervision centre was established to monitor the service in a manner similar to that described in Sect. 3.1 with basic functions of monitoring the fleet. Within SHOW, a booking platform was developed to enable the integration of the shuttle service within the local PT service. This integration highlighted that with a higher number of different software and hardware components over the various PT services in the whole region, the complexity increases. In other terms, throughout the work in SHOW, it became clear that replication of the AV service would be difficult in other cities or regions, because of variability of the ecosystem and lack of harmonization for the software used for dispatch as well as the supervision centre in its whole. Indeed, for now, the supervision centre is managed by the local operator at local level.

Rms recommends therefore an agreement of standardization of software and hardware for a supervision centre at regional level to facilitate scalability of the AV services (Fig. 5). Coordination through the regional PTA offers several advantages

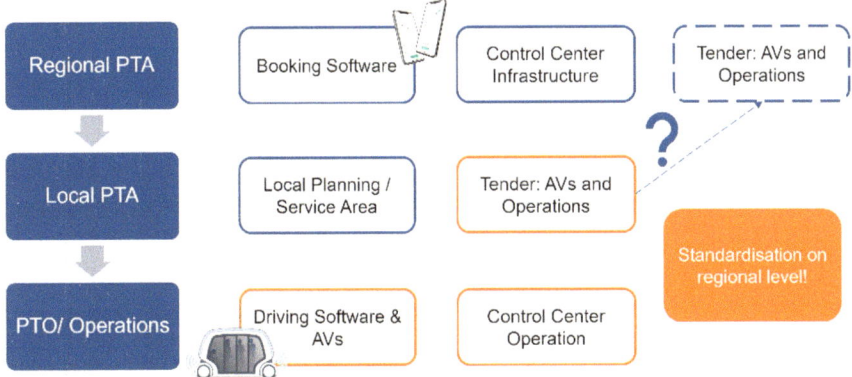

**Fig. 5** Vision of the governance model and the need for standardisation at regional level. *Source* Rms

for implementing comparable services across neighbouring cities: It removes barriers for passengers, who are unfamiliar with the interfaces, and it facilitates knowledge sharing among operators and city authorities regarding the new and complex technology involved. Mostly, standardization of the hardware and the interfaces between software systems (e.g., booking apps or operational software) as well as the centralization of automated control centres at regional level are highly recommended since it can streamline and expedite the rollout of automated on-demand services across the entire region, benefiting local authorities. The governance model depicted in Fig. 5 outlines a potential structure for this rollout process.

In this vision, operational software and control centres are organized at the regional level to ensure consistency in operational processes at the local level. Consequently, cities can focus on local planning without needing to independently acquire expertise on the new technology, particularly advantageous for smaller cities with limited capacity for such endeavours.

## 4 Discussion and Conclusion

In conclusion, while PTOs already possess the expertise needed to operate supervision centres for manually-driven fleets, there is a need to enhance these centres further considering automation challenges. This includes exploring the complexities of remote supervision, which present new safety and liability considerations. As the role of remote operators becomes increasingly important in supporting automated driving, it is essential to carefully consider safety measures, performance requirements, and key issues relevant to operators of remote vehicles.

The topic of disengagement of automated vehicles, particularly the handover to a safety-trained human operator, underscores the need for continuous learning from each situation and potential edge case (Khattak et al. 2020). Emergency handling procedures are crucial for operators, requiring comprehensive training to manage various emergencies effectively. Clear protocols and escalation procedures should be established and regularly practiced to ensure swift and efficient responses. Regulations, for instance in California[3] and in France,[4] support the notion that remote operators must be licensed to operate regular vehicles.

Furthermore, the current taxonomies of automated driving fail to acknowledge the potential for remote supervision (and potentially later driving) of AVs and the unique challenges faced by operators who monitor vehicles they are not physically occupying. Integrating the role of PTOs into industry-standard taxonomies is crucial for establishing regulatory frameworks related to training requirements, necessary equipment and technology, and a comprehensive inventory of potential use cases for remote supervision and operation (Mutzenich et al. 2021). The Human Factors in International Regulations for Automated Driving Systems (HF-IRADS) considers remote operation as a key priority for regulation and has called for the definition of automated driving to be broadened to include remote support (HF-IRADS 2020). This agreement on taxonomy will hopefully lead towards standardization of control centres. As presented in this chapter, the software used for dispatching and monitoring need harmonization so that cities and regions can easily replicate and scale the AV services from one place to another. All these aspects are critical considerations for future developments in remote supervision technology.

In conclusion, this chapter highlights challenges and potential enhancement that aim to facilitate the deployment of AVs as shared vehicles, seamlessly and safely integrated into public transport systems. This integration is crucial for maximizing the utilization of public transport and diminishing car usage and dependency. With a mature automated driving technology and advanced as well as standardized supervision centres, PTOs will be equipped to operate safely with reduced operational costs providing a higher quality of service and more attractiveness to the passengers.

**Acknowledgements** The work presented in this manuscript has been funded by the SHOW project which has been made possible by funding from the European Union's Horizon 2020 Research and Innovation Programme Under Grant Agreement no. 875530.

# References

Amador O, Aramrattana M, Vinel A (2022) A survey on remote operation of road vehicles. IEEE Access 10:130135–130154

---

[3] CA DMV, Adopted Regulation Title 13 Article 3.7 'Testing of Autonomous Vehicles'," 2019, https://www.dmv.ca.gov/portal/uploads/2020/06/Adopted-Regulatory-Text-2019.pdf.

[4] Ministère de la Transition Ecologique. Ordonnance n° 2021–443 du 14 avril 2021 relative au régime de responsabilité pénale applicable en cas de circulation d'un véhicule à délégation de conduite et à ses conditions d'utilisation https://www.legifrance.gouv.fr/eli/ordonance/2021/4/14/TRAT2034523R/jo/texte.

Cooke N (2006) Human factors of remotely operated vehicles. In: Human factors and ergonomics society 50th annual meeting, pp 166–169

HF-IRADS (2020) Human factors challenges of remote support and control. A position paper from HF-IRADS, Informal document submitted to UNECE GRVA-07-65 [Online]. Available: https://unece.org/fileadmin/DAM/trans/doc/2020/wp29grva/GRVA-07-65e.pdf

Hörcher D, Tirachini A (2021) A review of public transport economics. Econ Transp 25:100196 [Online]. Available: https://doi.org/10.1016/j.ecotra.2021.100196. ISSN: 2212-0122

Khattak ZH, Fontaine MD, Smith BL (2020) Exploratory investigation of disengagements and crashes in autonomous vehicles under mixed traffic: an endogenous switching regime framework. IEEE Trans Intell Transp Syst [Online]. Available: https://doi.org/10.1109/tits.2020.3003527

Mutzenich C, Durant S, Helman S et al (2021) Updating our understanding of situation awareness in relation to remote operators of autonomous vehicles. Cogn Res 6(1), 9 [Online]. Available: https://doi.org/10.1186/s41235-021-00271-8

Navetty Project (2022) European first in the Yvelines: autonomous electric shuttles running in a complex environment without an operator on board. Press Release, Les Mureaux, France. Available: https://www.transdev.com/en/press-release/autonomus-electric-shuttles-running-without-operator-on-board/

Ongel A, Loewer E, Roemer F, Sethuraman G, Chang F, Lienkamp M (2019) Economic assessment of autonomous electric microtransit vehicles. Sustainability 11(3):648 [Online]. Available: https://doi.org/10.3390/su11030648

Rhein-Main-Verkehrsverbund (RMV) (2022) Probefahrt in die Zukunft easy electric autonomous for you, Project brochure [Online, in German]. Available: https://www.probefahrt-zukunft.de/EASY_Brosch.pdf

Rhein-Main-Verkehrsverbund Servicegesellschaft mbH (rms) (2024) Autonome Mobilität, homepage [Online, in German]. Accessible: https://www.rms-consult.de/leistungen/new-mobility/autonome-mobilitaet/

SAE International (2021) Taxonomy and definitions for terms related to driving automation systems for on-road motor vehicles. J3016_202104, pp 41 [Online]. Available: https://doi.org/10.4271/J3016_202104

# Open and Modular Service-Oriented CCAM Architecture

Anastasia Bolovinou⑩, Georgios Spanos⑩, Antonios Lalas⑩,
Konstantinos Votis⑩, Emmanuel De Verdalle, Meriem Benyahya⑩,
Anastasija Collen⑩, Niels A. Nijdam⑩, Anna Antonakopoulou⑩,
Vassilis Sourlas⑩, and Maria Gkemou

**Abstract** Seamless integration of new types of mobility such as those realized by Connected Automated Vehicles (CAVs) into the existing public transport service requires new service-oriented orchestration platforms. In this work, we present a framework for automated mobility of passengers and goods, realized by the service-oriented architecture of the EU-funded research project SHOW (GA No 857730).

A. Bolovinou (✉) · A. Antonakopoulou · V. Sourlas
Institute of Communication and Computer Systems (ICCS), 9 Ir. Polytechneiou St., 15773 Zografou, Athens, Greece
e-mail: anastasia.bolovinou@iccs.gr

A. Antonakopoulou
e-mail: anna.antonakopoulou@iccs.gr

V. Sourlas
e-mail: v.sourlas@iccs.gr

G. Spanos · A. Lalas · K. Votis
Centre for Research and Technology Hellas, Information Technologies Institute, 6th Km Harilaou - Thermis Rd., 57001 Thessaloniki, Greece
e-mail: gspanos@iti.gr

A. Lalas
e-mail: alalas@iti.gr

K. Votis
e-mail: kvotis@iti.gr

E. De Verdalle
Information Technology for Public Transport a.i.s.b.l. (ITxPT), Rue Sainte-Marie 6, B-1080 Brussels, Belgium

Information Technology for Public Transport a.i.s.b.l. (ITxPT), Lindholmspiren 3-5, SE-402 78 Gothenburg, Sweden

E. De Verdalle
e-mail: emmanuel.de-verdalle@itxpt.org

M. Benyahya · A. Collen · N. A. Nijdam
Centre Universitaire d'Informatique, University of Geneva, Route de Drize 7, 1227 Carouge, Switzerland
e-mail: meriem.benyahya@unige.ch

H. Cornet and M. Gkemou (eds.), *Shared Mobility Revolution*, Lecture Notes in Mobility, https://doi.org/10.1007/978-3-031-71793-2_3

As part of the framework, intra-systems and cloud-to-everything interfaces are proposed, while high-level design specifications of architecture alternatives are critically reviewed. An actual example of how such an inclusive architecture is instantiated in Trikala SHOW pilot site for city-specific SHOW services deployment is presented. Finally, lessons learnt that are relevant to data access, interoperability and cybersecurity, based on the experience from all SHOW pilot sites, are outlined.

**Keywords** Connected automated vehicles · Mobility services · Cloud data management platform · Services-oriented architecture · Vehicle connectivity · Interoperability · Cybersecurity · Public transport · Freight

# 1  Introduction

As a part of the automotive industry changes focus from vertical, industry-based approaches, to delivering cross-domain services across multiple sectors, an expanding industry ecosystem is being created that includes vehicle manufacturers and their Tier 1 suppliers, cloud services providers, connected vehicle platform providers, independent software vendors and system integrators. All these actors need access to the data, interfaces and services offered by the connected vehicles and smart infrastructure.

Cooperative, connected, and automated mobility (CCAM) focuses on moving people and goods along the road networks in a safe, quick, cost-effective, comfortable and environmentally friendly manner using automated vehicles (AVs), leveraging Mobility-as-a-Service (MaaS) and Logistics-as-a-Service (LaaS) platforms. In SHOW, this involves transforming research pilot sites via numerous CCAM web-service platforms into fully operational service hubs, well-integrated into the existing local transit system, to ensure seamless availability and interoperability of transport and cargo CCAM services.

The technology supporting connected and semi or fully automated transport solutions has been rapidly evolving over the last few years. Connected vehicles and their corresponding backend and infrastructure communication systems are a reality in several cities, known as Cooperative—Intelligent Transportation Systems (C-ITS), while increased automation is on the horizon. Still, the implementation of C-ITS framework in urban road networks and in public transport (PT), as pursuit by the

A. Collen
e-mail: Anastasija.Collen@unige.ch

N. A. Nijdam
e-mail: niels@unige.ch

M. Gkemou
Centre for Research & Technology Hellas (CERTH), Hellenic Institute of Transport (HIT), 34 Ethnarchou Makariou St., 16341 Athens, Greece
e-mail: mgemou@certh.gr

well-established CEN TC 278 WG3 (CEN: CEN TC n.d.) working on road transport and traffic telematics, remains a challenge due to (a) lack of city authorities investments in smart infrastructure and (b) the specificities and complexity of the urban road traffic context, which is very different to the well-controlled motorway environment, and requires high accuracy in positioning, granularity in location referencing and continuous connectivity to enable cooperative services.

SHOW EU-funded project (SHOW n.d.) aims to support the deployment of shared, connected and electrified automation in urban transport, to advance sustainable urban mobility. Starting from the admission that there is no one-architecture-that-fits-all, the objective was to create a framework compliant with modern web-of-things architecture paradigm as shown in https://www.w3.org/TR/wot-architecture11/, Kovatsch et al. (2020) and based on international standards without imposing technology solutions. Based on data integration principle differentiations, three CCAM services-oriented flexible architecture variations that exhibit different manners of interoperability among the actors of the integrated system are derived, whilst cybersecurity mechanisms and communication protocols that apply vertically to all system layers are proposed. The ambition of this work is for its usage by public transport operators (PTOs) who wish to support CCAM services' deployment in their road network and can choose the architecture variation that fits better their business case.

This chapter is structured as following: Sect. 2 reviews the state-of-the art, Sect. 3 outlines the methodology adopted, Sect. 4 presents the proposed architecture through a conceptual, abstract functional and detailed functional view, Sect. 5 discusses CCAM architecture design choices with respect to data access, interoperability and cybersecurity aspects coming from SHOW project experience and beyond and finally, Sect. 6 concludes this chapter.

## 2   Relevant Initiatives and Selected International Standards

In order to support the integration of CAV/robot fleets in both mature and non-mature urban local eco-systems, a reference multitier architecture has to be conceptualized. The adopted by the EU CCAM research agenda (2021–2027) (CCAM Strategic Research and Innovation Agenda) acknowledges the need for inclusive new architectures that integrate seamlessly with existing systems: "Interaction with infrastructures, road and telecommunication infrastructure as well as automotive backend infrastructure, is crucial for the success of the system integration. Both elements of physical and digital infrastructure are important, in particular building a common understanding of what is required, how it can be achieved, and which roads should be prioritised with a view towards implementation." Taking into consideration that the PT network geography, topology and traffic have volatile characteristics and perhaps more importantly, being maintained by multiple organizations/authorities, the main objective was to design a system service-oriented architecture for urban PT context which is (a) compliant to regional used standards; (b) interoperable with existing systems; (c) sufficiently open to be easily adopted for most cities.

## 2.1 Architectures for CCAVs in PT and the CCAM Research Agenda

The SPACE (Shared Personalised Automated Connected vEhicles) project[1] launched in 2018 had the aim of placing public transport at the centre of the automated vehicles (AVs) revolution. SPACE has developed a high-level reference architecture that aims at ensuring a comprehensive and seamless integration of AVs with other IT systems in the mobility ecosystem using a fleet orchestration platform. The platform ensures a brand- and type-agnostic integration with the driverless vehicles and provides rich and open Application Program Interfaces (APIs) to develop professional and end users' applications while also being connected to PT backend systems and smart city system. It strongly influenced SHOW reference architecture derivation.

## 2.2 CCAM Services Enablers and the WoTs Paradigm

The automotive industry's transformation towards a seamless integration between cars as software on wheels and cloud ecosystems is under development. Ongoing discussions are giving rise to a growing industry consensus among OEMs, automotive suppliers and service providers: Common shared data models and APIs with standard features will enable companies to compete by providing customer value in a cost-effective way. A prominent initiative in this direction is the COVESA (Connected Vehicle Systems Alliance)/W3C alliance[2] that develops reference approaches for integrating operating systems and middleware present in connected vehicles and the associated cloud services. Formerly and under the auspices of the W3C automotive working group,[3] a Vehicle Data Interfaces Architecture was developed. This work was extended by the GENIVI alliance (former form of COVESA) for the promotion of new secure vehicle-cloud interfaces. They proposed the Secure Vehicle Interface (SVI) as a ready-to-deploy technology, based on three CEN/ISO standards namely the TS 21,177 (Intelligent transport systems ITS station security services for secure session establishment and authentication between trusted devices 2024), TS 21,185 (Intelligent transport systems Communication profiles for secure connections between trusted devices 2019) and TS 21,184 (Cooperative intelligent transport systems (C-ITS) Global transport data management (GTDM) framework 2021). SVI connects recognised and authorised external systems to the network within a vehicle. SVI then converts the vehicle manufacturer's proprietary vehicle data into a common language, which enables broad interoperability for competitive services irrespective of the manufacturer or brand of the vehicle. This is also aligned to the European Association of Automotive Suppliers (CLEPA) position represented by the work of

---

[1] https://www.uitp.org/projects/space/.

[2] https://covesa.global/about-covesa.

[3] https://www.w3.org/groups/wg/auto/.

ISO 20078 (Road vehicles Extended vehicle (ExVe) web services 2021) on Extended Vehicle (ExVe) specification.

The Web-of-Things (WoT) paradigm proposed by the World Wide Web Consortium (W3C) is an extension of the Internet-of-Things. There, the reference architecture is composed of three main layers which are (i) the connected device layercalled 'Things', (ii) the 'Gateway' and (iii) the 'Cloud' layer. Starting from th e bottom layer, the 'Things' layer, 'Things' could be physical or virtual. 'Things' are exposed as software objects with (restful) Application Programming Interfaces (APIs) by communicating events, properties and actions through popular protocols like HTTP, Web-sockets or MQTT. A very similar approach was followed in SHOW.

## 2.3   C-ITS Connectivity Relevant Aspects

Initially "silo-solutions" were predominantly developed and deployed for the different ITS service domains (e.g., eCall, Public Transport, Traffic and Traveller Information). Still the last decade, the concept of C-ITS that includes support for sharing of data, components and software (e.g., radio transceivers, localization equipment, software-based facilities) amongst service domains has gained consensus in EU leading to the development and release of numerous C-ITS standards by ISO, CEN and ETSI working groups, which promote hybrid communications, neutrality of technology where applicable, portability of ITS applications, security and privacy. As shown in the European ITS Platform white paper (Eu EIP 2016), improvements for selected services have become feasible with cross-domain connectivity, such as Safety Related Traffic Information (SRTI) or Real Time Traffic Information (RTTI), see Commission Delegated Regulations (EU) No 886/2013 (EU n.d) and 2015/962 (Eu). Additionally, the access to (near) real time incident data enables a variety of new services which have not been feasible before like Overtaking Warning, Extended Intersection Collision Warning, as example of what envisioned by C2C consortium.[4]

## 2.4   PT Data Access for 3rd Party Service Providers and NAPs

Moving towards a Single European Transport Area requires a digital layer interlinking all of the elements of transport. Building up this digital architecture involves open and common standards and interfaces and an efficient, but secure data ecosystem. This is why Member States are setting up their National Access Points (NAP)[5] to facilitate access, easy exchange and reuse of transport related data, in order to help support the provision of EU-wide interoperable travel and traffic services to end users. NAP is a European intermediary platform and it is part of EU ITS Directive

---

[4] https://www.car-2-car.org/about-C-its#c133.

[5] EC-ITS / NAPs https://ec.europa.eu/transport/themes/its/road/action_plan/nap_en.

2010/40/EU (EU 2010) specification. All delegated regulations supplementing the ITS Directive refer to certain standards to be used when exchanging information with NAPs. While DATEX II is prevalent, the NeTEx CEN/TS 16,614 (Public transport—Network and Timetable Exchange (NeTEx) 2020) and SIRI CEN/TS 15,531 (SIRI) standards are also stated. "Transmodel", is the short name for the European Standard "Public Transport Reference Data Model" (EN 12,896) (Public transport—Reference data model). The ITxPT association[6] specifies communication protocols and hardware interfaces to offer a full interoperability of IT systems (on-board and back office) for PT applications based on CEN/TC 278 (https://www.itsstandards.eu). This is the standardization body that manages the preparation of standards in the field of ITS in Europe. Within this, WG3 defines ITS standard for Public Transport.

## 3 Design Principles and Methodology

The architecture work is recommended to be started with interviewing the local ecosystem actors in each deployment site so that local authorities and PT operators would be involved from the beginning and that the proposed architecture is able to accommodate and adapt to local constrains and existing legacy systems under operation. Following this logic, we ensure that the cloud-based CCAM data management platform will be able to efficiently integrate with existing fleet management and PT backend systems. People and freight transport services can be considered.

The proposed architecture supports a set of on-board and operational backend intelligent applications. In SHOW, it served as a harmonized and supervised design framework to be used by the SHOW pilot sites for integration of their local subsystems into the new CCAM service-oriented paradigm. In this way, while a big degree of flexibility has been given to the local teams on how to design their local ecosystem for SHOW fleet piloting activity, a minimum set of design requirements was ensured to be followed by all, entailing the following aspects:

- Common data format and data access design principles for both static and dynamic content (via standardized interfaces)
- Interoperable data exchange among heterogeneous data providers (maximizing standardized interfaces)
- Harmonized integration of external data sources through APIs
- Data privacy and cyber-security cross-layers mechanisms recommendations and integrated such mechanisms in the SHOW Mobility Data Platform.

The methodology followed in order to derive the reference architecture included the following steps:

1. Based on the project use cases (SHOW 2021a) and the knowledge acquired via the interviews with the majority of the SHOW pilot sites technical boards, the

---

[6] https://itxpt.org/specifications/.

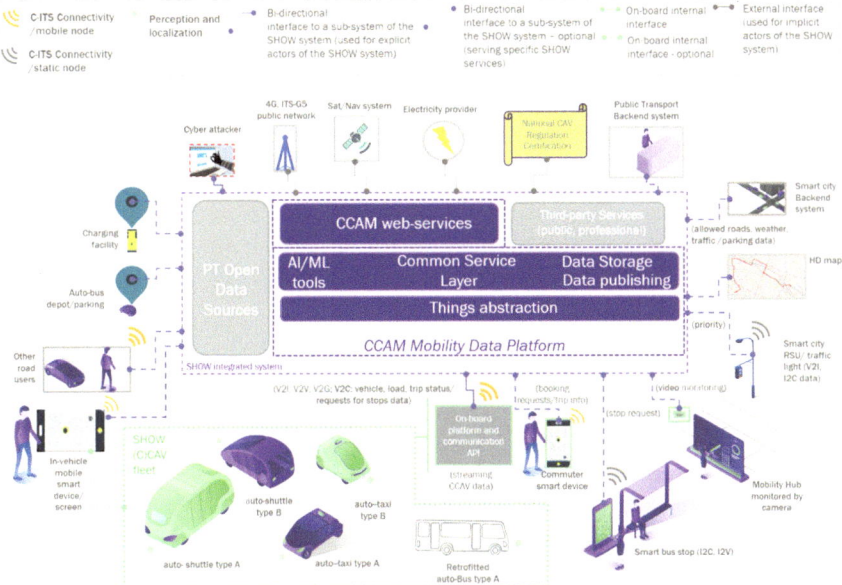

**Fig. 1** CCAM service-oriented architecture conceptual and abstract functional view: actors and type of data exchanged in the SHOW integrated system. *Source* SHOW D4.1 (SHOW 2021b)

conceptual architecture was defined and mapped to CCAM value chain stakeholders (see Figs. 1 and 2). In parallel, all internal and external subsystems interacting in each SHOW pilot site were identified.

2. Combining the SPACE reference architecture with the WoTs architecture paradigm, a service-oriented architecture for CCAM services is proposed.

3. Based on the C4 nested model[7] for creating architecture diagrams, the work was split into providing the three views of the system architecture: conceptual, abstract functional and functional.

## 4 CCAM Service-Oriented Architecture Views

The proposed CCAM contextual architecture is presented in Fig. 1. It models the attributes of and the interaction among the CCAM ecosystem actors and the CCAM Mobility Data Platform in an integrated system. CCAM actors include: AV operators, PT operators, riders, other road users, public authorities, 3rd party data providers, 3rd party services providers, automakers and legislative bodies. In Fig. 2 each component is further mapped to stakeholders to facilitate external re-use of the architecture. The integrated system includes a physical layer of connected 'Things' (including vehicles, infrastructure and other road users connected to the cloud) and

---

[7] https://c4model.com/.

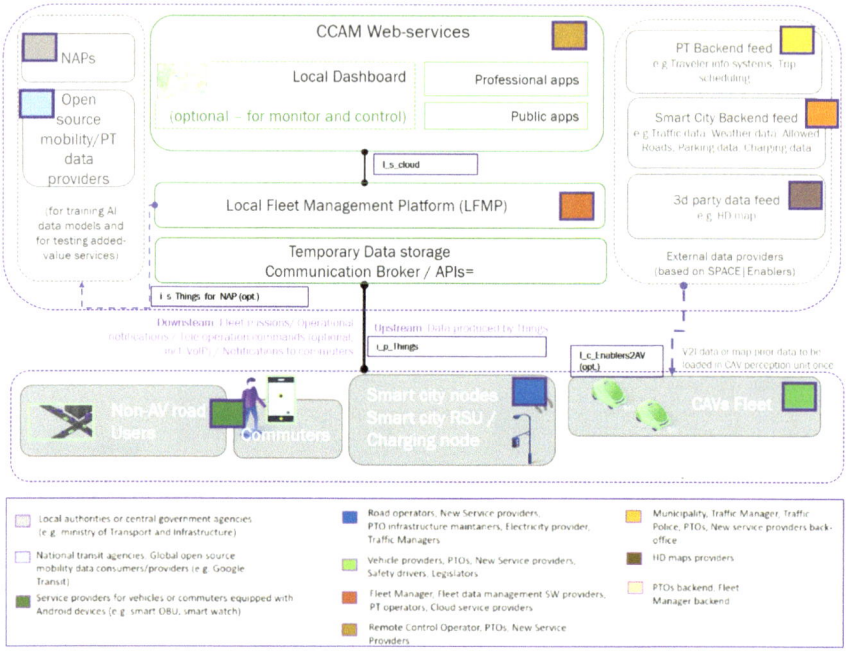

**Fig. 2** CCAM service-oriented architecture functional view, plus main components mapped to stakeholders. *Source* Own, authors—based on SHOW D4.4 (SHOW 2023)

a cloud layer (Local Fllet Management Platform) hosting processes of data ingestion, handling and dissemination. Both layers are allowed to act as data providers and data consumers via a bi-directional link between the two. The higher part of the cloud layer hosts the web services implemented on top of the Local Fllet Management Platform that can be either public or commercial ones. In this architecture, the Dashboard is also considered as a dedicated web-service. External data providers are depicted on the left and on the right parts of the cloud layer in the diagram of Fig. 2.

In Fig. 3, a more detailed architecture with intra layers components specified is proposed, in which:

- Accounting for more flexibility in data sharing practices, two variations of the CCAM reference architecture are proposed. Depending on the availability of real-time streaming data, a third variation is added. In I_p_Things interface '_p' denotes proprietary, while in the I_s_cloud interface '_s' denotes standardized and in I_c_enablers2AV, '_c' denotes standardized or custom interface from external provider. All nodes within the physical layer may be connected to the cloud via the I_p_Things or I_s_Things interfaces. Additionally, they may be interconnected within the same layer they belong into, via short-range Vehicle-to-Everything (V2X) ad-hoc networks. A physical Vehicle-to-Grid (V2G) interface provided for charging the electric CAVs is also considered.

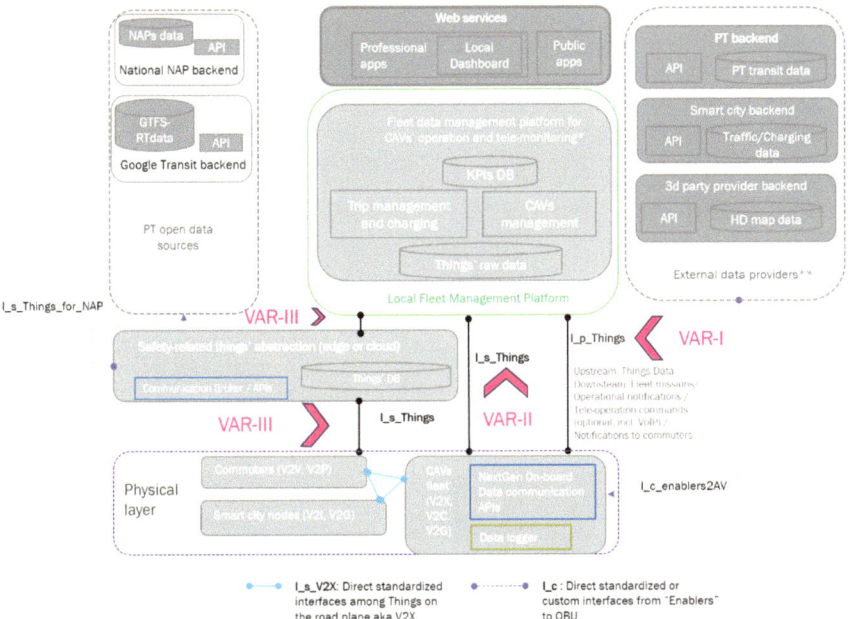

**Fig. 3** CCAM service-oriented architecture variations that assumes multiple CCAM data ingestion cloud platforms for open real-time data publication to enable safety related services (VAR-I, VAR-II and VAR-III denote the three architecture variations). *Source* Own, authors

- The Things' data ingestion and data publisher is denoted as discrete layer (Things' abstraction) to unify various operations performed on raw data in modern cloud data sharing platforms like data normalization, filtering, anonymization, authentication, etc. Support for various data feed rates e.g. per millisecond, seconds, trip, day implies the support of IoT event-driven architectures. It is called **cloud** Things' abstraction to differentiate from the possibility of a similar layer located on the **edge** (of the physical layer). However, in the future where 5G will be more widespread, this gateway that also collects the data from all connected entities, could be indeed physically implemented as an edge component.

Figure 4 presents how this reference architecture is applied to support the SHOW CCAM deployment at Trikala pilot site (as described in SHOW 2022). As observed therein, the Trikala Fleet Management Platform (TFMP) includes an interface to an external cloud data provider, i.e., the Traffic Management Center (TMC). On top of the TFMP, apart from the SHOW On-Demand Transport (DTR) app, two other services are integrated namely the Traffic Light Signaling Monitoring, Control and Management (TLSMCM) service and the Remote Control Center (RCC) service. Both make use of the I_p_things interface which is used for bi-direcitonal data exchange between the CAV or infrastructure and the cloud-based TFMP (for more details on exchanged content please see the figure).

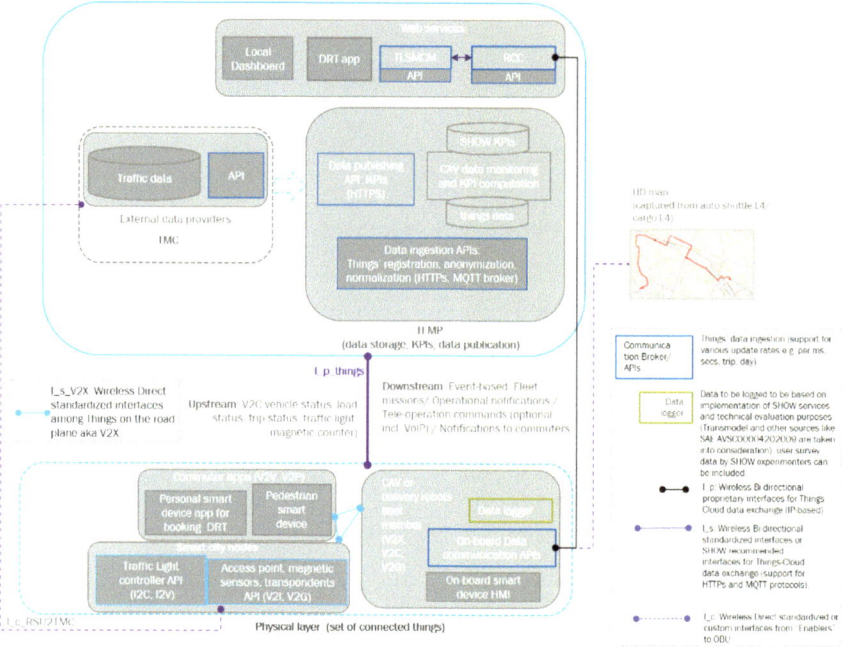

**Fig. 4** The example of trikala local fleet management platform architecture that follows CCAM service-oriented architecture variation no. 1. *Source* SHOW D4.3 (SHOW 2022)

## 5 Proposed Architecture Variations and Important Aspects

Enhancing integration and interoperability of CCAM ecosystems heavily depends on (a) data openness (b) data interoperable exchange and (c) secure and standardized interfaces within the proposed services-oriented CCAM architecture, as well as between the proposed architecture and the legacy systems operated by smart city and local PT operators. In the following, we discuss derived lessons learnt for the two above aspects based on this work and SHOW pilots' design experience.

### 5.1 Data Streaming and Access

Next generation CAV on-board architecture is expected to support real time streaming of vehicle generated data in a secure and efficient way that does not affect the CAVs' internal communication channels (security and privacy-preserving vehicle operation system). Building upon MaaS business models, the need for next generation vehicle platforms and increased openness in data sharing strategies have recently started to be shaped. The distinct characteristics of the three architecture variations proposed in Sect. 4 (see Fig. 3) are discussed below:

- Variation—I: CCAM services deployment based on peer-to-peer agreements between CAVs' owners, CAV operators and 3rd party service providers. Indirect access to Things' data, subset of data available via cloud- to cloud file transfers or ideally via pub/sub APIs (assumes an agreement with project CAV data owners and operators); CCAM non time-critical services can be offered. Most of the SHOW pilot demos adopted this variation which respect local city CCAV providers and PTO's constrains.
- Variation—II: Both direct and indirect access to Things' data in multiple update rates via two data ingestion cloud platforms. Data ingestion and publishing layer should follow an event-driven architecture that supports real time streaming of data. CCAM non time- critical services can be offered more openly. Few of the SHOW pilot demos adopted this improved variation.
- Variation—III: (adopting an equal data access approach for 3rd party service providers on an agreed minimum set of CAV data) Safety critical CCAVs services deployment based on minimum set of shared CCAV data. This is aligned with the FAIR data principles promoted by the EU. It enables fleet data to be streamed real time via an intermediary vendor-neutral server. Next generation AV on-board architecture is assumed that supports real-time communication from in-vehicle gateways. CCAM time-critical services related to safety can be offered. The authors recommend the adoption of this variation as more future-proof and B2B-friendly at European level and believe that this shall be subject of future regulation by the EU.

## 5.2   Connectivity and Interoperability

Based on reviewing the SHOW pilot sites CCAM architecture instantiations, a few general remarks can be drawn:

- Adoption of EU data formats for PT services (e.g. SIRI, NeTEx), as proposed by ITxPT specifications, varied per site, mainly based on CCAM services maturity. Towards the future standardization of CCAM deployment for PT we would make these two additional notes:
  a. Vehicle data: It is important that the access to vehicle data through EU standards based on PTA/PTO requirements data is supported. If such data are requested by PT operator shall then be made available by OEM either as historic data or real-time feed. This is part of ongoing discussion in the frame of EU data act (Proposal for a Regulation of the European Parliament and of the Council on harmonised rules on fair access to and use of data).
  b. Mobility data: this is related to NAP (National Access Point) and MMTIS delegated regulation (Multimodal travel information services) which request to publish mobility data (static one today and dynamic one in coming revision of MMTIS) using EU standards.

- Local Fleet Data platform is often operated by the OEMs (i.e. shuttles' providers) and hence the fleet to cloud data format and protocols are proprietary.
- When connected mobility is deployed via connectivity to infrastructure, other vehicles or other road users (V2X), ETSI compliant C-ITS protocols are typically used.
- For the realization of the real time connection between the fleet and the cloud, MQTT and WebSockets mechanisms were recommended and followed by all.

## 5.3 Cybersecurity

SHOW focuses on the cybersecurity aspects for CCAV services deployment in PT scenarios. For this purpose, relevant security-critical parameters had been identified and the special characteristics of the automated driving functions and the provided infrastructure as well as the CCAV services deployed on top, had been taken into account, mainly focusing on securing the SHOW core cloud components and the communication of the SHOW set of connected things to this cloud backend. This work considered the following standards: ISO/SAE 21,434 (Road vehicles—Cybersecurity engineering 2021), ISO 31000 (Risk management), ISO 26262 (Road vehicles—Functional safety), SAE J3061 (Cybersecurity Guidebook for Cyber-Physical Vehicle Systems) for joint safety and security engineering as well as J3101 (Hardware Protected Security for Ground Vehicles) for cybersecurity risk management.

At the level of the SHOW cloud data portal, the project makes use of advanced mechanisms for detection of cyber-attacks through novel tools with the aim to cover wide aspects of cyber security anomaly detection and intrusion detection. SHOW cybersecurity threat analysis and risk assessment and selection of mitigation strategies have been dealt in SHOW (D4.1 SHOW 2021b, D4.3 SHOW 2022 and D4.4 SHOW 2023) as well as cyber-security defence mechanisms implemented and tested (incl. a network-based Intrusion Detection System) (SHOW). In each pilot site, additional cyber security mechanisms may apply depending on the cybersecurity protocols of the vehicle owners and the existing cloud infrastructure components/APIs. In order to assist the local sites cybersecurity planning, the project also circulated a cybersecurity best practices document related to automotive applications (for more details see SHOW 2023).

From the analysis of the cybersecurity and privacy questionnaire reported in SHOW (2023) complementing the cybersecurity monitoring work in SHOW, it was deduced that (a) there is not a common privacy strategy for all the pilot sites, however they all respect the GDPR (General Data Protection Regulation) and FDIC (Federal Deposit Insurance Act) obeying to the privacy obligations stemming from these; (b) although all the pilot sites utilize secure communication protocols for data transmission and make use of Secure Onboard Communication in order to be protected against crucial cyber-attacks such as Jamming and Spoofing, no common cybersecurity strategy applies on either physical or cloud layers as this depends on the specificities of each deployment; and (c) pilot sites that deploy V2X connectivity make use of standardized ETSI C-ITS protocol and specifications which support cybersecurity; and d) use of vehicle-to-cloud for remote operation/monitoring requires special care with respect to connectivity redundancy and cybersecurity.

# 6 Conclusion

This work proposed a framework for seamless integration of CAVs into the existing and established public transport paradigms, following a service-oriented orchestration approach. Since many of the cities already operate PT management systems, significant effort has been put to propose generic interfaces considering them, and derive design recommendations for all systems and actors that need to be interfaced and interact, within the augmented CCAM service-oriented ecosystem, such as public transport operators, CAVs' fleet operators, infrastructure operators or third party service providers. For this purpose and respecting each local CCAM ecosystem constrains, three design alternatives with respect to CCAM data sharing practices are proposed as blueprint in order to assist practitioners and stakeholders involved to select what suits them better. A future-proof architecture variation was proposed (variation no. 3) that enables fleet data to be streamed real time via an intermediary vendor-neutral server which in turn enables CCAM time-critical services offer.

As demonstrated by the SHOW piloting activity and at the level of the local sites' cloud infrastructure, a big variety exists among the architectures deployed in the SHOW pilot sites, based on their maturity and commercial components' integration as well as the CCAM use cases under focus. There is no one architecture that fits all solutions and hence adoption of each architecture variation shall be assessed on local and context-specific manner. Important aspects like interoperability, cybersecurity and data access that should be supported by a CCAM architecture have been evaluated. Adoption of EU data formats for PT services as well as of European C-ITS standards can greatly aid towards this direction and can lead to more mature CCAM services that will ideally support a more "plug&play" services' offer. Still, there is a standardization gap for newly emerging use cases including remote fleet monitoring and bilateral communication with the fleet in case of an emergency that shall be covered for facilitating future safe deployment of CCAVs in urban PT. CCAVs fleet management systems' maturity shall be assessed by national authorities with respect to remote operation/monitoring capabilities that require special care in terms of real-time data exchange for reaction to time-critical events, connectivity redundancy for continuity of services and cybersecurity to avoid external disturbances.

This work provided a general framework for a service-oriented CCAM architecture and described the main interfaces that need to be in place for orchestrating different connected data providers and consumers. In-vehicle architectures that support real-time communications and vehicle-to-cloud interfaces, considered as part of future Software-Defined-Vehicles' deployment, have not been evaluated from this work and it is left for future research.

**Acknowledgements** The research is part of the SHOW project that has received funding from European Union's Horizon 2020 research and innovation programme under grant agreement no. 875530.

# References

CCAM Strategic Research and Innovation Agenda, Version 1.0, 02/11/2020 [b9.2]

CEN: CEN TC 278 WG3. Road transport and traffic telematics–Public Transport (PT)

Cooperative intelligent transport systems (C-ITS) Global transport data management (GTDM) framework (2021) ISO/TS 21184:2021

Cybersecurity Guidebook for Cyber-Physical Vehicle Systems (2016) SAE J3061_201601

EU EIP (2016) Interoperability requirements for C-ITS services with infrastructure involvement, Deliverable 1 of sub-activity 4.4 (Cooperative ITS Services Deployment Support), Version 1.0 [Online]. Available: https://www.its-platform.eu/wp-content/uploads/ITS-Platform/Achiev ementsDocuments/IntegratingC-ITS/EU%20EIP-44-D1-C-ITS%20Interoperability%20Requ irements-v1.0.pdf

EU (2013) Commission delegated regulation No 886/2013. Available: https://eur-lex.europa.eu/ legal-content/EN/ALL/?uri=celex%3A32013R0886

EU (2015) Commission delegated regulation No 2015/962. Available: https://eur-lex.europa.eu/ legal-content/EN/TXT/PDF/?uri=CELEX:32015R0962&from=EN

EU (2010) Directive 2010/40/EU. Available: https://eur-lex.europa.eu/eli/dir/2010/40/oj

Hardware Protected Security for Ground Vehicles (2020) SAE J3101_202002

https://www.itsstandards.eu

Intelligent transport systems Communication profiles for secure connections between trusted devices (2023) ISO/TS 21185:2019

Intelligent transport systems ITS station security services for secure session establishment and authentication between trusted devices (2024) ISO 21177:2024

Kovatsch M, Matsukura R, Lagally M, Kawaguchi T, Toumura K, Kajimoto K (2020) Web of Things (WoT) Architecture. Available online: https://www.w3.org/TR/wot-architecture/

Multimodal travel information services. https://eur-lex.europa.eu/eli/reg_del/2017/1926/oj

https://www.w3.org/TR/wot-architecture11/ (Note: The W3C WoT architecture is designed to describe what exists, and only prescribes new mechanisms when necessary)

Proposal for a Regulation of the European Parliament and of the Council on harmonised rules on fair access to and use of data (Data Act). https://eur-lex.europa.eu/legal-content/EN/TXT/?uri= COM%3A2022%3A68%3AFIN

Public transport–Network and Timetable Exchange (NeTEx) (2020) CEN TS 16614

Public transport—reference data model, CSN EN 12896

Risk management (2018) ISO 31000

Road vehicles—Functional safety (2018) ISO 26262-1

Road vehicles Extended vehicle (ExVe) web services (2021) ISO 20078-1:2021

Road vehicles—Cybersecurity engineering (2021) ISO/SAE 21434:2021

SHOW (2021) D5.1: big data collection platform and data management portal". deliverable of the horizon-2020 SHOW project, Grant Agreement No. 875530

SHOW (2021a) D1.2: SHOW use cases, Deliverable of the Horizon-2020 SHOW project, Grant Agreement No. 875530 [Online]. Available: https://show-project.eu/wp-content/uploads/2021/ 04/SHOW-WP01-D-UIP-002-02_-_SHOW_D1.2_SHOW_Use_Cases_SUBMITTED.pdf

SHOW (2021b) D4.1: Open modular system architecture and tools–first version (November 2021 revision), Deliverable of the Horizon-2020 SHOW project, Grant Agreement No. 875530 [Online]. Available: https://show-project.eu/wp-content/uploads/2022/10/D4.1-Open-modular-system-architecture-and-tools-first-version.pdf

SHOW (2022) D4.3: Open modular system architecture and tools–second version (2022). Deliverable of the Horizon-2020 SHOW project, Grant Agreement No. 875530 [Online]. Available: https://show-project.eu/wp-content/uploads/2022/03/SHOW_D4.3_system_architecture_second_version_final.pdf

SHOW (2023) D4.4: Open modular system architecture and tools–third version. Deliverable of the Horizon-2020 SHOW project, Grant Agreement No. 875530 [Online]. Available: https://show-project.eu/wp-content/uploads/2023/03/SHOW-WP04-D-UIP-004-01_-_SHOW_D4.4_Open_modular_system_architecture-third_version_Final.pdf

SHOW project, Grant Agreement No. 875530, https://show-project.eu/

SIRI–Service interface for real-time information relating to public transport operations, CEN: EN 15531

# Infrastructure and Operations

# Physical and Digital Infrastructure (PDI) Support for Automated Vehicles—Case Studies in Austria

**Andreas Hula, Martin Dirnwöber, Stefan Ladstätter, Andrea Schaub, Patrick Luley, Daniel Tötzl, Stephan Wittmann, Dominik Schallauer, Isabela Erdelean, Karl Rehrl, Alexander Frötscher, Bernhard Monschiebl, and Veronika Prändl-Zika**

**Abstract** Within the EU Horizon 2020 project SHOW (GA No 875530) auxiliary measures, via evaluating and adapting physical road infrastructure (for instance lane markings, traffic signs, sight distances), as well as via digital support were explored for their potential contribution to enabling automated shared mobility services on the road environment. This line of research was then followed up on in the EU Horizon

A. Hula (✉) · A. Schaub · S. Wittmann · I. Erdelean · V. Prändl-Zika
AIT Austrian Institute of Technology, Giefinggasse 2, 1210 Vienna, Austria
e-mail: andreas.hula@ait.ac.at

A. Schaub
e-mail: andrea.schaub@ait.ac.at

S. Wittmann
e-mail: Stephan.wittmann@ait.ac.at

I. Erdelean
e-mail: Isabela.Erdelean@ait.ac.at

V. Prändl-Zika
e-mail: Veronika.Praendl-Zika@ait.ac.at

M. Dirnwöber · D. Schallauer · A. Frötscher · B. Monschiebl
AustriaTech, Raimundgasse 1/6, 1020 Vienna, Austria
e-mail: Martin.Dirnwoeber@austriatech.at

D. Schallauer
e-mail: Dominik.Schallauer@austriatech.at

A. Frötscher
e-mail: alexander.froetscher@austriatech.at

B. Monschiebl
e-mail: Bernhard.monschiebl@austriatech.at

S. Ladstätter · P. Luley
Joanneum Research, Steyrergasse 17, 8010 Graz, Austria
e-mail: Stefan.Ladstaetter@joanneum.at

P. Luley
e-mail: patrick.luley@joanneum.at

© The Author(s) 2025                                                                                    51
H. Cornet and M. Gkemou (eds.), *Shared Mobility Revolution*, Lecture Notes
in Mobility, https://doi.org/10.1007/978-3-031-71793-2_4

Europe Project AUGMENTED CCAM (GA 101,069,717), tackling more specifically with PDI support for automated mobility. This chapter presents the activities and findings of these two projects in relation to physical infrastructure adaptations for automated vehicles for two pilot sites in Austria.

**Keywords** Connected Automated Vehicles (CAVs) · Digital Dynamic Maps (DDM) · Physical and Digital Infrastructure (PDI) · Cooperative Intelligent Transportation Systems (C-ITS)

# 1 Introduction

Enabling automated mobility (AM) has been at the forefront of applied research and technological development in the mobility domain, as fully automated vehicles (AVs) promise benefits to systemwide optimisation of mobility systems, through high vehicle standards, optimal driving behaviour (Milakis 2019; Taiebat et al. 2018) and perhaps central guidance through (cooperative) intelligent transport systems ((C)-ITS).

However, present day efforts struggle to deliver on these goals, as AVs on the road still need to gain widespread acceptance (Becker and Axhausen 2017). In particular, challenging interactions (conflicts or collisions) with other road users, are still an ongoing concern, for building trust with this new technology (Zhang et al. 2019).

Approaches to remedy these shortcomings consist, among others, of trying to improve sensor systems and the formal representation of human behaviour on the road, to the point that automated systems will rarely cause or get involved in crashes. Such an approach is challenging, as the rules driving humans' traffic behaviour, both as motorized and unmotorized road users, are not explicit and not necessarily known to even the actors themselves. Identifying these factors through modelling and data analysis is just as difficult to reliably validate. Hence, another path is to aim for adaptations of infrastructure, be they digital (for instance by providing detailed real-time updated digital maps) or physical (for instance high quality road markings and road signs well readable for the automated systems). These can be summarized in physical and digital infrastructure (PDI) measures (Cucor et al. 2022). The advantage of PDI is that physical infrastructure is essential for conventional vehicles as well and the PDI building on this can, once in place, be utilized by all suitably equipped vehicles, to different extents. The downside of PDI approaches is that they require infrastructure adaptation or expensive set-ups and typically need to provide their

D. Tötzl
Yunex Traffic, Siemensstraße 90, 1210 Wien, Austria
e-mail: Daniel.toetzl@yunextraffic.com

K. Rehrl
Salzburg Research Forschungsgesellschaft, Jakob Haringer Straße 5/3, 5020 Salzburg, Austria
e-mail: karl.rehrl@salzburgresearch.at

benefits in a less tailor-made way for the individual vehicle compared to sensors on a specific vehicle.

The Horizon 2020 project SHOW (SHOW Project) sparked an evaluation of a set of physical and digital infrastructure measures (SHOW; Erdelean et al. 2023) and this has led to follow up work in the Horizon Europe project "AUGMENTED CCAM" (ACCAM) (Augmented CCAM Project), focusing explicitly on the PDI support that can be beneficial for vehicles of no or different level of automation.

Whether to tackle any issue of enabling for Cooperative Connected Automated Mobility (CCAM) through PDI or onboard vehicle sensors is a fundamentally different approach to solving the challenges of automation for vehicles. It is worth noting that these two approaches are not mutually exclusive; implying rather that the PDI approach can be a means to support early penetration with potential benefits to AVs of lower SAE levels (SAE International 2021)[1] or in specific Operational Design Domains (ODDs) in mixed traffic conditions, whilst the in-vehicle sensor-based approach, along with remote supervision advancements, is striving for the fully driverless operability under any real-traffic condition. To date vehicles equipped with an extensive array of sensors still rely on additional existing high definition (HD) maps, which can be considered elements of PDI (see for instance Waymo's operations described in Waymo (2024), which explicitly mentions the mapping out of the area of operation).

While the implementation of PDI measures for CCAM is still in its early stages, the relevant experiences on two pilot sites in Austria (SHOW-Project) are presented in the current chapter. The role of PDI in supporting AVs to extend their ODD is discussed.

## 2 Background

Systematic work on PDI support has been carried out in the AUGMENTED CCAM (ACCAM) project (Augmented CCAM Project), a key outcome of which has been the harmonized and prioritised PDI support classification schema for CCAM (to be published as "Deliverable D2.1", at (Augmented CCAM Project), by the project).

The ACCAM PDI support classification schema consists of 81 elements of support for CCAM (and beyond, as it distinctively addresses all key cohorts of vehicles, including conventional vehicles) that have been further ranked based on four (4) criteria, namely: (a) benefit to traffic safety, (b) benefit to traffic efficiency, (c) benefit to enabling automated driving and (d) costs. This has enabled a prioritisation of a long list of PDI potential adaptations, on the basis of road authorities' and road operators' needs and focus areas (for instance suitability for application on highways, in rural areas and in urban areas, as well as the categories "Traffic Control/

---

[1] Refers to the Society of Automotive Engineers (SAE) J3016 standard, that defines six levels of automation, from 0 to 5, source: 2024 SAE International (2021).

Management", "Road infrastructure", "Weather conditions", "Illumination conditions", "Road pavement/ Surface conditions", "Traffic conditions", "Surrounding traffic", "Digital services", "Communication enablers" and "Digital enablers").

Based on the state of implementation at the pilot sites two (2) PDI support elements have been chosen for review in two pilot sites in Austria (Koppl in Salzburg and Klagenfurt in Carinthia), as follows:

- HD map of physical road features, which relates to digital dynamic maps (DDMs) discussed in SHOW (see Jomrich 2020, Hula et al. 2023).
- PDI support solutions for protection of vulnerable road users (VRUs) and safe interaction during lane change and other manoeuvres.

In Austria, SHOW pilots for automated driving employed automated shuttles (AS) that require a detailed representation of the road environment and a strong Global Navigation Satellite System (GNSS) connectivity, to operate on fixed routes in the road environment. A human operator had to be present for safety reasons and speeds were limited for test use, also to conform with the national regulatory frameworks. In relation to the selected PDI support elements mentioned above, it was found that:

- The shuttle's systems utilized HD mapping. Information on road dimensions (lane widths, curve radius) were encoded in the HD map, but no particular infrastructure adaptations beyond trimming vegetation were in place during operation (both for Koppl in Salzburg and Klagenfurt, Carinthia).
- For the protection of VRUs and the support of vehicle manoeuvres in critical areas, a camera and C-ITS based solution was implemented at one of the pilot sites (deployed in Klagenfurt, Carinthia).

The subsequent chapters detail the implementation of PDI at the two sites, starting with a high precision digital map in Koppl, which allows an AS to navigate safely in a specified physical environment. This is followed by the case of a Yunex "awareAI" camera-based solution for ensuring safe interactions between an AS and (nonconnected) vulnerable road users at an intersection. The final section discusses these cases in a wider context.


## 3   Ultra HD Maps—The Case of Koppl, Salzburg, Austria

The first pilot site to provide insight into a PDI measure and the setup thereof was Koppl near the City of Salzburg in Austria (see also SHOW Project). The municipality of Koppl is located in the peri-urban area of the City of Salzburg. The route links the centre of Koppl municipality to the "Sperrbrücke" bus stop, which acts as an intermodal interchange where passengers are able to transfer from the AS bus to the public bus line. Including start and terminus stops, the route serves four bus stops in each direction. The rural location of this route presents a number of challenges (e.g., partly lacking points of reference such as buildings or road markings for the reliable positioning of the vehicle), poor GNSS signal quality, and rudimentary road infrastructure without signalized intersections.

The vehicle deployed here was a small shuttle for up to six (6) passengers, with a human driver being present at all times, in case a take over from the automated system was required. The vehicle could move in automation mode on fixed routes with the potential to stop at a number of fixed stops. Thus, a high-quality digital representation of the road network in question was important, as the vehicle would plan its manoeuvres on the road lanes according to this digital representation. This approach should simulate an SAE level 4 operation though the fixed routes were a notable limitation.

This pilot site served to showcase how the HD maps can support the operation of AVs and the challenges in setting it up. In fact, in this case an ultra-high definition (UHD) map was created, with an accuracy in the range of a few centimetres.

The UHD map was created for in-vehicle use by means of (1) mobile mapping, (2) semi-automatic vector data extraction and (3) assigning semantic metadata. The mobile mapping process was performed with a Leica Pegasus 2 Ultimate 3D laser scanning system which is car-mountable, provides dual profiler LiDAR scanning (Light Detection and Ranging, see for instance Guan et al. 2016) up to 2 million points a second and a 360° panoramic camera with 24-megapixel resolution. For typical measurement ranges of up to 50 m distance to the carrier car, an accuracy of at least 5 mm can be expected. Global positioning is provided by a triple band GNSS receiver and a tactical grade inertial measurement unit (IMU) sensor, resulting in an absolute localization accuracy of the produced point cloud of 2 cm after processing with base station correction data. The measurement system is illustrated along with a colorized measured point cloud in the following Fig. 1.

The 3D point cloud serves as ground truth for extracting vector information, which provides the basis for the digital UHD map. This map is composed of the following attributes: reference line, road markings, lane segments until the curb, outlines of sign plates and signals including metadata, upper edges of barriers, walls and fences and stop lines. Given proper features identifiable in the point cloud and/

**Fig. 1** Leica Pegasus 2 ultimate mobile mapping system (left), Colorized 3d point cloud (right). *Source* Joanneum Research

**Fig. 2.** 3D vector annotations overlaid with point cloud (left), 3D vector annotations without point cloud (right). *Source* Joanneum Research

or measurement images (e.g., intensity variations, geometric attributes, texture) the extraction of vector data can be performed semi-automatically (Fig. 2).

For assigning semantic metadata, such as a topological connection graph, manual modifications are still necessary though. A seemingly simple intersection can consist of multiple conditional turn-relations (see Fig. 3).

Particularly in rural areas, connection arms may lead into very narrow roads without markings, making automatic topology connections challenging. Also, one way roads have to be taken into account when constructing the connection graph. In addition, for the AV to reliably detect right of way of intersecting traffic, stop and yield signalling has to be (1) assigned correctly to the intersecting road and (2) to the correct ground marking stop line. While this task is solvable through spatial search algorithms, a final manual quality control is still necessary (see also Rehrl et al. 2022 for another discussion of the map making process).

This in-vehicle map would allow for the safe operation of the AV through local-ization of the AV on the correct lane and by providing the possible connections. A challenge can arise from the static nature of the map. If essential features (like

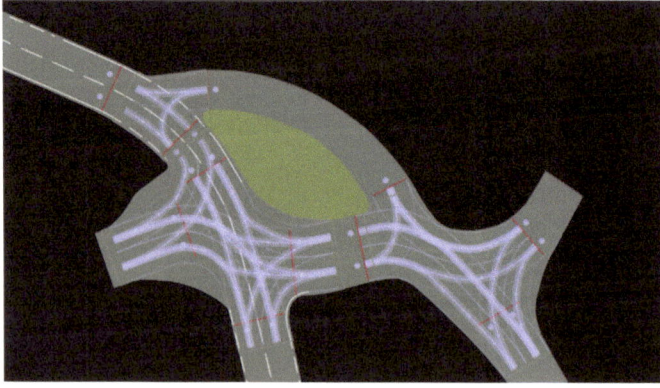

**Fig. 3** Example of possible turning relations in the village centre of Koppl. *Source* Joanneum Research

available lanes) were to be blocked (for instance because of construction works) the vehicle would need different manoeuvring rules and lane recognition to adapt to this situation. Ideally, such information would be provided by the actors having the authority to impact roads in this manner i.e. by the transport authorities already in the process of permitting, for instance, road works and sharing that information with the map providers in advance and frequently. Of course, this places additional burdens on the relevant authorities, so technical support for such data provision would be helpful. PDI support for such situations (road works and the inherent change in road topology, as well as challenges to vehicle sensors detecting this situation) is being discussed, for instance in ACCAM (Augmented CCAM Project).

The main advantage of an HD map is, that it can allow for the vehicle to be operated without needing physical infrastructure adaptations. In the particular case of an UHD map, the vehicle would have a precise (down to a few centimetres) representation of the driving lane due to the ultra-high definition and could plan its manoeuvres safely and reliably.

## 4 PDI Support for AV-VRU Interactions at a Crossing in Klagenfurt, Carinthia, Austria

In this section, a support measure for shuttles' interactions with other (in particular nonconnected) road users through digital infrastructure, as was implemented at the intended Carinthia pilot site (see SHOW-Project), is described. The pilot site is in an urban environment, connecting the train and bus station of Klagenfurt West with the university of Klagenfurt, a science park, a work hub and residential area, etc. with an AS. Conditions of mixed traffic, public roads and high traffic demand. Over 4 km route with 14 stops and travel time of 30 min (3 routes) (1 route—3 metro modes (2, 3, 4 km length)).

The vehicle deployed in Klagenfurt was a shuttle for up to 12 passengers, with a safety driver being present at all times, in case a take over from the automated system was required. A HD map was used to enable the vehicle to travel along fixed routes and reach fixed potential shuttle stops. The vehicle is designed and labelled as a prototype with SAE Level 4 capabilities, although its current operation is restricted to a fixed set of routes.

The goal in the urban context of Klagenfurt, was to compensate through PDI for limitations inherent to the AVs, like the limited range of vehicle sensors, and allow for safe interactions between them and other (also not-connected) road users, in particular unmotorized vulnerable road users like pedestrians or cyclists.

A PDI support solution was deployed in a critical junction of the Carinthian pilot site (see Fig. 4 below). To ensure a safe passage through the pedestrian crossing, an AS requires knowledge of an unobstructed crosswalk to prevent collisions. Therefore, infrastructure-based support was implemented, featuring the following elements:

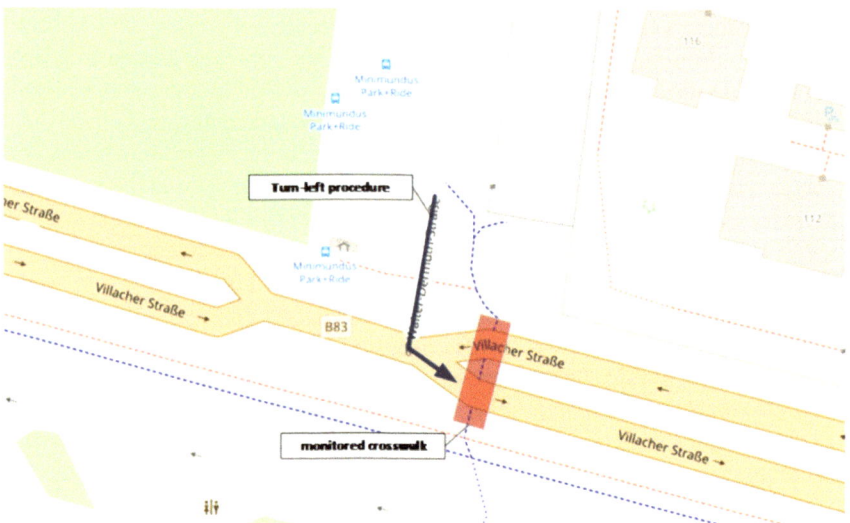

**Fig. 4** Klagenfurt test scenario, highlighting the turn-left procedure of the shuttle and the intersecting crosswalk. *Source* Modified from Open Street Map

- A camera system with on edge AI processing ("awareAI", Traffic and awareAI 2024) for detecting vulnerable road users (VRUs) on the crosswalk. The awareAI camera has been deployed on top of an existing traffic light pole at a height of approximately 6 m. This installation expended the camera's field of view, thereby ensuring the acquisition of high-quality video data from the surveilled areas. The awareAI core, functioning in real-time, undertakes the processing of incoming video stream data. Throughout the implementation process, two specific detection zones were established, corresponding to the possible directions of travel All detection tasks are performed within the local processing unit of the camera system, guaranteeing the highest level of data protection as only fully anonymized information is processed by external systems. Figure 5 illustrates the detection zone configuration of the monitored area, highlighting the two segments of the crosswalk (represented as red layout) as well as the corresponding waiting areas (represented as yellow layout). In the event of a successful trigger activation, the awareAI core system generates the respective C-ITS information which is updated and triggered per second.
- A C-ITS compliant roadside unit (RSU) as part of the aware AI set-up responsible to broadcast warning messages in the event of detected individuals from the camera system. C-ITS ETSI (ETSI standards) standardised messages have been implemented and broadcasted by the RSU. The current vehicle status is broadcasted by vehicles using CAMs (Cooperative Awareness Message), information on traffic lights is provided via SPATEM/MAPEM (signal phase and timing message, information on the layout) and information on events is disseminated

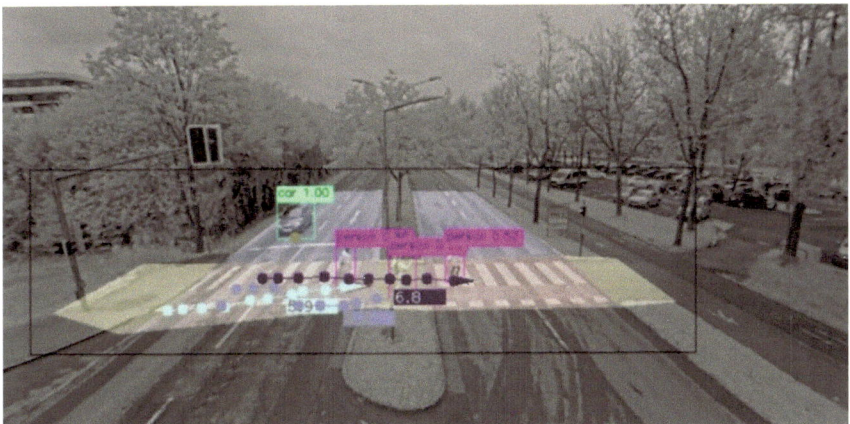

**Fig. 5** Illustration of the awareAI detection parameters. *Source* Yunex Traffic Austria GmbH

via DENM (Decentralized Environmental Notification Messages). See Fig. 6 for a visualisation of an interaction.

- Additionally, a HD-map of the whole pilot site was set up, to support automated navigation. The map functions similarly (similar degree of detail and support of operation) as in the above case described for Koppl and is also static, with the messaging described above used to identify relevant dynamic situations, but not having no impact on the digital representation of the road infrastructure itself.

The detection zone was exclusively configured to trigger when pedestrians or cyclists were located in the corresponding zones. Other categorized entities, such as vehicles, were detected within zones as well, but did not initiate any C-ITS information transmission.

A technical validation test was performed to assess the reliability of C-ITS messages. In this test case the test system tested the situation statically while parked in a parking space close by and also dynamically while driving towards the crosswalk with pedestrians crossing. The test was conducted in the timeframe of an hour and the triggering condition was intentionally prompted about 5 times with several additional triggers caused by pedestrians that were in the area coincidentally. The conclusion of the technical validation test was very positive as every time a person walked across the crosswalk a DENM was sent out, which would be sufficient to alarm a vehicle about the pedestrian presence.

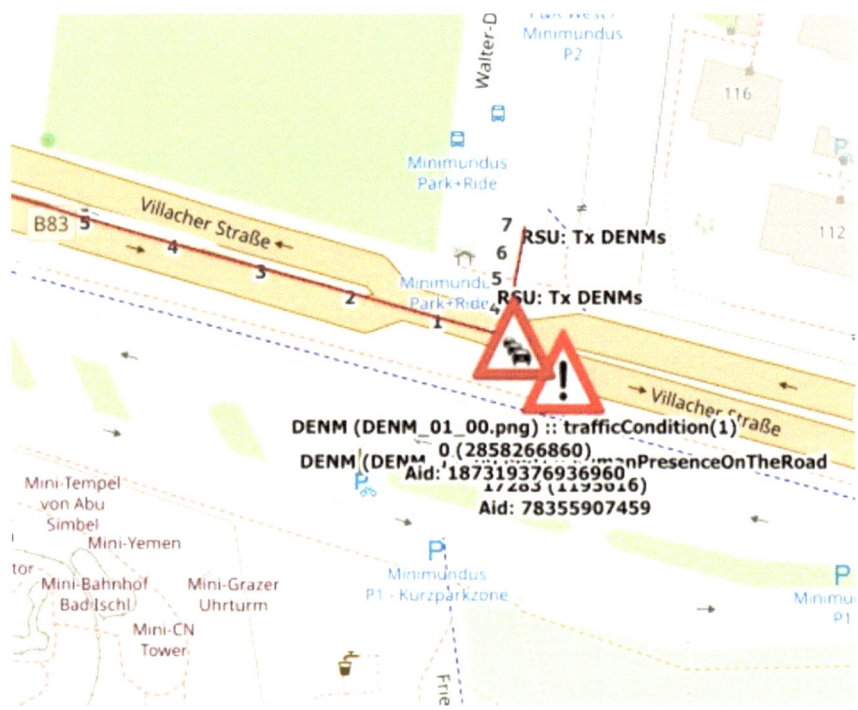

**Fig. 6** Warning messages of cars and pedestrians crossing, issued and broadcasted upon detection by the awareAI based system. *Source* Modified from OpenStreetMap

## 5 Conclusions

This work presents how specific PDI support elements have been deployed by two Austrian pilot sites to support the operation of ASs and extend their ODD in a rural and urban setting respectively. The first method was a detailed high-definition map provided to the AV to safely determine lanes and the domain of operation in general. HD maps (sometimes UHD maps, if high precision estimates are desired) is an essential tool for current AVs and the process of setting them up is increasingly optimized (see SHOW Project, Jomrich 2020, Hula et al. 2023, Rehrl et al. 2022). One of the main advantages of using an HD map is that, ideally, no physical infrastructure adaptation is necessary to provide a sensor-equipped AV with all it needs to navigate. We note however that this promise might fail due to sudden changes in the infrastructure, which would require substantial operational adaptations by the vehicle. The second was a real-time AI based object recognition system, warning an AV about potential collisions with VRUs. The system was tested by project partners and correctly issued warnings during pedestrian presence in the crossing. An approach such as this addresses the ongoing concerns and debate on how to best integrate (in particular unmotorized) VRUs in a connected environment (see Scholliers et al. 2017; Ess et al.

2021). This approach is challenging with regards to cost-benefits, as deploying such a system on every crossing would increase maintenance needs and costs.

In conclusion, digital measures like maps are more common than physical adaptations, which seems rational, given that digital measures leave less of an impact on the existing infrastructure and thus face lower hurdles to deploy once the technology for setting them up quickly is available.

The implementation of the use case described above in the pilot site in Carinthia highlighted how AVs can be supported by the PDI enabled solutions and, thus, make operation safer, protecting especially VRUs, which might not be connected to any digital systems (i.e. might not carry a mobile phone or other sensor) reliably. The advantage of this pilot site-based approach is the relatively safe deployment and the potential of implementing supporting PDI in advance and only moving towards more natural driving conditions in a step-by-step manner. In contrast, while several US companies have shifted from testing to commercial deployments, they often seek to establish their case in largely unsupported everyday environments. This reflects a strategic preference to minimize reliance on physical infrastructure, which they may have limited control over, and adapt to varied real-world scenarios. The two approaches could naturally support each other, if data were to be shared between respective operators, improving the safety and speed of deployment for all. To achieve this synergy, alignment of public and private interest is required.

This underscores the potential of PDI solutions to enhance road safety and emphasizes the importance of further research and development in this area for realizing safer and more inclusive autonomous transportation systems. The range of PDI measures under evaluation is being expanded in research projects like ACCAM, considering also the challenges found here (for instance dynamic changes in road topology due to road works). Improvements in publicly managed (C)-ITS and digital representation in the public domain might encourage data sharing and help tackle the challenges faces by the early pilots described here.

**Acknowledgements** SHOW has received funding by the European Union's Horizon 2020 research and innovation programme under Grant Agreement No. 875530. The AUGMENTED CCAM Project has received funding from the European Union's Horizon Europe programme under Grant Agreement No. 101069717.

# References

Augmented CCAM Project. Accessed 12 Oct 2023 from https://augmentedccam.com/
Becker F, Axhausen KW (2017) Literature review on surveys investigating the acceptance of automated vehicles. Transportation 44:1293–1306
Cucor B, Petrov T, Kamencay P, Pourhashem G, Dado M (2022) Physical and digital infrastructure readiness index for connected and automated vehicles. Sensors 22(19)
Erdelean IP, Schaub A, Stefan C, Vanzura M, Prändl-Zika V, Hula A (2023) Assessment of physical road infrastructure to support automated vehicles in an urban environment. Transp Res Procedia 72:2385–2392

Ess J, Luppin J, Antov D (2021) Estimating the potential of a warning system preventing road accidents at pedestrian crossings. LogForum 17(3):441–452. https://doi.org/10.17270/J.LOG. 2021.605

ETSI standards, Retrieved May 4th from https://www.etsi.org/

Guan H, Li J, Cao S, Yu Y (2016) Use of mobile LiDAR in road information inventory: a review. Int J Image Data Fusion 7(3):219–242. https://doi.org/10.1080/19479832.2016.1188860

Hula A, Schaub A, Stefan C, Prändl-Zika V, Erdelean IP (2023) Applying a semi-automated workflow for digital dynamic maps to support the operation of automated vehicles. Transp Res Procedia 72:735–742

Jomrich F (2020) Dynamic maps for highly automated driving—generation, distribution and provision. Dissertation, Darmstadt, Technische Universitat, 10.25534˙/tuprints-00009702

Milakis D (2019) Long-term implications of automated vehicles: an introduction. Transp Rev 39(Jg., Nr. 1, S):1–8

Rehrl K, Groechenig S, Piribauer T, Spielhofer R, Weissensteiner P (2022) Towards a standardized workflow for creating high-definition maps for highly automated shuttles. J Location Based Serv 16(2):119–151

SAE International (2021) SAE levels of driving automation refines for clarity and international audience. https://www.sae.org/blog/sae-j3016-update

Scholliers J, van Sambeek M, Moerman K (2017) Integration of vulnerable road users in cooperative ITS systems. Eur Transp Res Rev 9:15. https://doi.org/10.1007/s12544-017-0230-3

SHOW Project (n.d.) Accessed 12 Oct 2023 from https://show-project.eu/

SHOW Project, SHOW Deliverable D9.3 Pilot experimental plans, KPIs definition and impact assessment framework for final demonstration round. Accessed 31 Jan 2024 from https://show-project.eu/wp-content/uploads/2022/10/D9.3-Pilot-experimental-plans-KPIs-definition-imp act-assessment-framework-for-final-demonstration-round.pdf

SHOW-Project, Austrian Megasites overview. Accessed 12 Oct 2023 from https://show-project.eu/mega-sites-austria/

SHOW Project (n.d.) SHOW deliverable 8.1. Accessed 12 Oct 2023 from https://show-project.eu/wp-content/uploads/2022/10/D8.1-Criteria-catalogue-and-solutions-to-assess-and-improve-physical-road-infrastructure.pdf

Taiebat M, Brown AL, Safford HR, Qu S, Xu M (2018) A review on energy, environmental, and sustainability implications of connected and automated vehicles. Environ Sci Technol. https://doi.org/10.1021/acs.est.8b0012

Yunex Traffic (2024) Aware AI. Accessed 16 April 2024 from https://www.yunextraffic.com/awareai/

Waymo (2024) Waymo driver. Accessed 16 April 2024 from https://waymo.com/waymo-driver

Zhang T, Tao D, Qu X, Zhang X, Lin R, Zhang W (2019) The roles of initial trust and perceived risk in public's acceptance of automated vehicles. Transp Res Part C: Emerg Technol 98

# Real-Life Automated Public Transport Operations at the SHOW German Mega Site—Experiences and Lessons Learned

**Yun-Pang Flötteröd, Katharina Karnahl, Sofia Pavlakis, Anja Holdermüller, Sven Ochs, Marc René Zofka, Kai Dietl, and Mitja Mook**

**Abstract** Automated shuttles promise a solution to achieve sustainable mobility and meet the manpower issue in public transport, and are expected to extend public transport service coverage and improve riding comfort. The piloting activities at three German mega sites successfully demonstrated fixed and non-fixed mobility services

Y.-P. Flötteröd (✉)
Institute of Transportation Systems at the German Aerospace Center, Rutherfordstrasse 2, 12489 Berlin, Germany
e-mail: yun-pang.floetteroed@dlr.de

K. Karnahl
Institute of Transportation Systems at the German Aerospace Center, Lilienthalpl. 7, 38108 Braunschweig, Germany
e-mail: katharina.karnahl@dlr.de

S. Pavlakis
Rhein-Main-Verkehrsverbund Servicegesellschaft mbH (Rms), Am Hauptbahnhof 6, 60329 Frankfurt Am Main, Germany
e-mail: sofia.pavlakis@rms-consult.de

A. Holdermüller
Bahnen der Stadt Monheim GmbH (BSM), Daimlerstrasse 10A, 40789 Monheim am Rhein, Germany
e-mail: A.Holdermueller@bahnen-monheim.de

S. Ochs · M. R. Zofka
FZI Research Center for Information Technology, Haid-und-Neu-Str. 10–14, 76131 Karlsruhe, Germany
e-mail: ochs@fzi.de

M. R. Zofka
e-mail: zofka@fzi.de

K. Dietl
TraffiQ Lokale Nahverkehrsgesellschaft - Frankfurt Am Main mbH, Stiftstrasse 9-17, 60313 Frankfurt Am Main, Germany
e-mail: k.dietl@traffiQ.de

M. Mook
Ioki GmbH, An Der Welle 3, 60322 Frankfurt Am Main, Germany
e-mail: mitja.mook@ioki.com

and the technical possibilities to let shuttles move without a pre-defined virtual track and cope with a ride-booking application. Results show that the shuttles operated with promising performance. In total, in the EU project SHOW (GA No 875530), approximately 107,000 km and 40,000 passengers were respectively travelled and transported by twelve (12) automated shuttles. This chapter explicates the technical preparation, shares the collected experiences and proposes recommendations.

**Keywords** Automated shuttles · Demand responsive transport · Automated public transport service · On-demand mobility

# 1 Introduction

Rapid advancements in sensor technology and data processing within the automated automotive field enable vehicles to behave in an intelligent manner. Automated shuttles (AS) are one of the resultant products. This enables public transport (PT) operators to enhance their mobility services to meet passengers' needs and to implement overall changes as public mobility is facing growing manpower and skills shortages (European Commission 2023). Many areas across Germany, e.g. Berlin, Hamburg, Frankfurt, Karlsruhe, Friedrichshafen and Munich, are dedicated to improving PT systems and services with AS, and have over the past years conducted various real living labs and test sites to explore the performance of AS-based mobility solutions (VDV 2024; Connected Automated Driving Europe 2024). Several test beds for automated driving have been set up as well (Federal Ministry for Digital and Transport 2024; DLR 2024; Federal Highway Research Institute 2024; Test Area Autonomous Driving Baden-Württemberg 2024). In addition, the AFGBV (Ordinance on the approval and operation of motor vehicles with autonomous driving functions in specified operating areas) (Betriebsbereichen der Justiz 2022) has been launched and the automated driving regulation is unified for entire Germany with some exceptions since 2023. Before that, AS permission was issued by the respective district government and safety report was inspected by the respective regional TÜV (technical supervisory association). AS applicability and the operation system stability of AS mobility services continuously need to be examined and verified in practice. In this sense, different mobility services were proposed, demonstrated, and evaluated within three (3) cities, constituting the German mega site of the European project SHOW[1]: Frankfurt-am-Main, Monheim am Rhein and Karlsruhe (SHOW 2024).

The pilot activities focused on both fixed and DRT (Demand Responsive Transport) oriented mobility services. At all sites, AS were considered an important component for meeting the agreement of mobility and climate protection in future PT. This is since AS can be efficiently used in PT, especially for DRT services, so that they only run upon request. This avoids empty kilometres and makes PT more flexible (Dytckov

---

[1] https://show-project.eu/.

et al. 2022). Especially in areas with little or no PT, automated DRT systems can be an important addition. A driver-based DRT service is already available in Frankfurt and in other cities and communities within the area of the local public transport authority (PTA) Rhein-Main-Verkehrsverbund (RMV). To expand the successful service in the RMV region, the plan is to replace some vehicles by driverless AS.

Similar to the Rhine-Main area, the Monheim city administration has set its goal to transform Monheim into a smart city, with AS being an integral part of this scheme. Step by step AS are being integrated into the local PT system. The federal state Baden-Württemberg, where the Karlsruhe site is located, also considers AS an improvement to the PT system and various funding programs have been made available to build the required competence for automated PT. The following sections introduce each site with its technical preparation and integration, summarize the experiences collected, propose recommendations and conclude with prospects.

## 2 Overview of the German Mega Site

The German mega site considers both fixed and DRT mobility services. Each site is described in detail below.

### 2.1 Frankfurt Site

The RMV—as the regional PTA in the Rhine-Main-Region—has worked together with its 100% consulting-subsidiary rms on automated PT systems since 2019. Together with the vehicle manufacturer EasyMile (EM), the software provider ioki and local partners, the automatization of existing DRT-services has been further investigated within the SHOW (https://show-project.eu/) project. The services were operated by the local PTA traffiQ and the PT company Stadtwerke Verkehrsgesellschaft Frankfurt am Main (VGF).

Between November 2022 and October 2023, an additional service connected a neighbourhood with limited access to local PT—Frankfurt's Riederwald district—to the nearby underground station Schäfflestrasse, creating real added value for residents. The service was free of charge, but passengers had to book their trips via a smartphone app. While the local population is socio-demographically quite diverse, particularly elderly residents made use of the service. Operating hours ranged from Monday to Saturday between 8 a.m. and 3 p.m. with no fixed schedule and flexible booking via the DRT app. The deployed vehicles were two EM shuttles EZ10 Gen3 shuttles, with which the routes were digitally mapped before the start of operation. A safety operator was always on board to fulfil legal requirements.

The service area covered a large part of the Riederwald district and connected the residential area to Frankfurt's central PT system as well as to other Points of Interest (POIs), such as a supermarket, a pharmacy and restaurants shown in Fig. 1a.

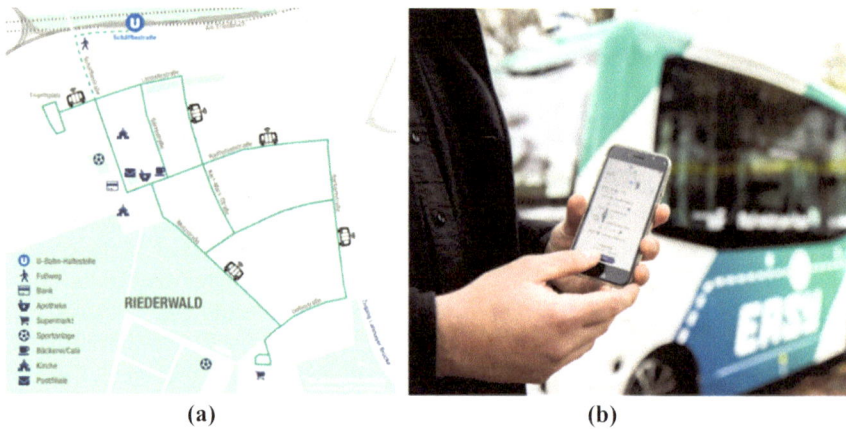

**(a)**                                                    **(b)**

**Fig. 1** Service area and booking app at the Frankfurt site: **a** service area and route; **b** booking app "RMV EASY". *Sources* RMV and Christof Mattes

In order to serve as many people as possible and still ensure access to the stops, more than 30 virtual stops were spread across the service area, guaranteeing a short walking distance to the subsequent pick-up point. A virtual stop is a non-fixed (as in flexible, unestablished and movable) place where vehicles are able to briefly stop in order to pick-up or drop-off passengers (Armellini et al. 2021). At eight of the 30 virtual stops, barrier-free access for wheelchairs and rollators was provided with an electronically extendable ramp, integrated in the shuttles. All roads in the area have a medium traffic volume consisting of cars, pedestrians and bikes and a speed limit of 30 km/h. There are one-way and two-way lanes as well as pedestrian crossings on site. Parking is mostly clearly structured so that it would not impede the shuttle's path.

The main technical objective of this site was to smoothly connect two software solutions, i.e. the booking app in Fig. 1b and the automated driving software (ADS). The architecture of the software connection is described in Sect. 3.1. The second objective was to test two artificial intelligence (AI) systems which are to enhance service quality in future operations when no safety operator is available: an AI-based camera system by T-Systems which detects specific situations inside the shuttle and communicates critical situations to the passengers, and a voice bot for passenger communication, serving for receiving and providing passengers' answers regarding PT-connections. Both systems are further described in Sect. 3.2.

## 2.2 Karlsruhe Site

The Karlsruhe site was composed of two modified EM shuttles EZ10 Gen2 shuttles for PT-oriented automated driving. The FZI Research Institute for Information

Center developed the required software and operated the shuttles. These vehicles collectively showcased a variety of applications, spanning from first and last-mile passenger transportation to cargo transport. The pilot activities took place in the city of Karlsruhe, Germany. The vision was to optimize the ride AS as much as possible in regard to speed, comfort and safety and to test the potential of innovative vehicle concepts. Also explored was the potential of DRT to replace underutilized buses in suburban areas. The shuttles were previously modified to move freely according to traffic conditions (including for instance double-parked vehicles on the roadway), rather than adhering to a pre-defined virtual track. The modifications included mounting additional six LiDAR sensors for better perception of the environment. Still, the main difference and implementation novelty has been related to the safety and underlying architecture of the automated driving function (Ochs et al. 2023; Lambing et al. 2021). The system was primarily validated by using an overall system-oriented fleet test under real conditions in the Test Area Autonomous Driving Baden-Württemberg (Federal Highway Research Institute 2024). The ioki app was adopted to facilitate smooth online-booking by the passengers.

The pilot site encompassed two distinct areas (see Fig. 2). The first area was KIT Campus East: a controlled environment with minimal traffic, ideal for testing automated driving functions. Particularly during the initial stages of system integration, the KIT Campus East offered a 5000 m$^2$ open space for dynamic tests and experimental driving trials. The second area was Weiherfeld-Dammerstock, a semi-urban district in the southwest of Germany. This setting involved interactions with pedestrians, cyclists and e-scooter riders. Multiple schools in the vicinity added complexity to this area, as this could lead to critical scenarios, where children could be more at risk. PT operations were conducted in this area from December 2022 to September 2023. Passengers varied from tourists to research scientists to the majority of local residents in the target area. The area was deliberately selected for its diverse demographic makeup, providing valuable insights into the utilization of DRT-based PT. The last-mile aspect was also emphasized, as the tram station serving the target area is readily accessible, with the farthest point being just two kilometres away. This served as a prominent example of the last-mile use case. Additionally, in both areas, there was smart infrastructure available, facilitating remote monitoring of the vehicles. Through V2X interfaces, intersections exchanged data with vehicles and vice versa, which were then visualized at the FZI control centre (see Fig. 3).

## 2.3   Monheim Site

The PT operator Bahnen der Stadt Monheim (BSM) started automated driving in February 2020, acting on the decision of the Monheim city council in December 2017. Monheim is a town with a population of around 45,000 residents in the Rhine region between Cologne and Düsseldorf. The target group for AS are residents (mainly elderly citizens) and tourists. The line A01 connects Monheim's old town with the central bus station Monheim Mitte and is 1.7 km long (see Fig. 4). The AS must

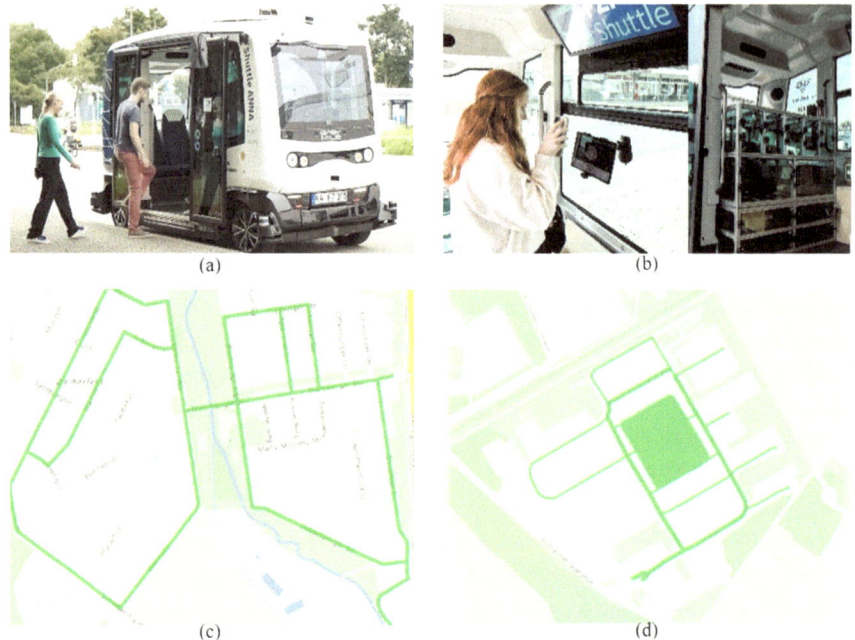

**Fig. 2** Illustration examples at the Karlsruhe test site: **a** shuttle on the way; **b** on-board layout; **c** route in Weiherfeld-Dammerstock with OpenStreetMap (2024); **d** route in KIT Campus East with OpenStreetMap (2024). *Source* Authors' own pictures

**Fig. 3** Layout illustration of the FZI control centre. *Source* Authors' own pictures

**Fig. 4** Service area and shuttle operation at the Monheim site: left: the AS route; right: AS at the bus terminal. *Sources* BSM and Tim Kögler

master narrow roads, a pedestrian zone and even a passageway through an old tower. The fleet consists of five EM shuttles EZ10 Gen2 vehicles, complemented by three (3) additional EM shuttles EZ10 Gen3 vehicles since 2023. There are eight (8) fixed stops along the route. Line A01 is completely integrated into Monheim's PT system and operates seven days a week from 9 a.m. until 9 p.m. More than 30 specially trained safety operators work to provide this special service. Figure 4 shows the AS service area and the AS operation situation at the bus terminal.

## 3 Summary of the Pilot Sites

Table 1 shows the overall performance and the composition of the SHOW (https://show-project.eu/) German sites. In total, approximate 107,000 km and 40,000 passengers were respectively travelled and transported by 12 AS, 2 of which moved freely at the Karlsruhe site, whilst the others 10 in Monheim and Frankfurt sites followed a pre-defined virtual track. In Monheim, the shuttles have been running not only on residential roads, but also in a shared space with cyclists and pedestrians. Regarding mobility service, both the Frankfurt and Karlsruhe sites provided free DRT services via the ioki booking app. By contrast, the AS service at the Monheim site is a paid service for non-residents and has been integrated into the regional PT system. The respective ticketing price corresponds to the public transport tariff of the federal state of North Rhine-Westphalia, i.e. 3.3 EUR (adult) and 1.9 EUR (child) within a single tariff zone (VRR 2024). People with a day/group ticket or a PT ticket subscription can also use the AS. In addition to passenger transport, the Karlsruhe site also offered parcel transport service, and the Frankfurt site provided additional on-board information during their AS services for enhanced safety and comfort.

**Table 1** Key data of the SHOW German mega site for PT operation

| Key data | Frankfurt | Karlsruhe | Monheim |
|---|---|---|---|
| Shuttles | 2 EM⁻Gen 3 | 2 modified EM-Gen2 | 5 EM-Gen2<br>3 EM-Gen3 |
| Maximum permitted speed | 20 km/h | 20 km/h | 18 km/h |
| Operating logic | Virtual track | Move freely | Virtual track |
| Environment | Residential road | Residential road, KIT campus | Residential road, shared space |
| Route length | 2.7 km | 4.5 km | 2.7/1.7 km (before/after 11.2022) |
| Service type | DRT with 30 + virtual stops | DRT with 22 virtual stops | Fixed schedule and stops |
| Operator | VGF | FZI | BSM |
| Operation period | Nov. 2022–Oct. 2023 | Dec. 2022–Dec. 2023 | May 2022–Dec. 2023 |
| Operation schedule | 8–15 on weekdays | 9–16 on Fridays and weekends | 9–21 daily |
| Booking software | Ioki | Ioki | Local PT system |
| PT integration | No | No | Yes |
| Riding fare | Free | Free | Non-residents; (3.3/1.9 EUR for adult and child respectively) |
| Key audience | Residents | Residents | Residents/tourists |
| Other services in/ given by AS | AI-camera system, voice bot | Parcel transport | N/A |

## 4 Technical Preparation and Integration

To carry out the above-mentioned services, the respective technical components had to be prepared, enhanced and connected to each other as explained below.

### 4.1 DRT Booking System and Connection to AS

For the SHOW site in Frankfurt, EM provided the AS platform; the booking software and underlying technology platform were provided by ioki both for the Frankfurt and the Karlsruhe sites. Building on their experience from previous and on-going joint automated projects (FZI 2024a, b), one focus was on further testing the interaction and communication of the two software systems. In this way, specific requirements associated with automated DRT services have been identified. Two examples are as follows: (a) it was tested how the shuttles handle ride requests in an automated way, and (b) investigated which specific requirements arise from the operation of virtual stops for automated DRT services.

Ioki's user-friendly ride pooling platform includes the app "RMV EASY" and the intelligent algorithm for routing and trip bundling. The user interface for passengers is an application where customers can book and cancel journeys and specify special requirements, e. g. taking along a rolling walker or wheelchair. In addition, passengers receive information about the estimated departure and arrival times (Mook and Pavlakis 2023).

All relevant information for automated PT converges on the ioki platform, which coordinates the shuttles by efficiently bundling the trip requests and setting clearly defined "missions" that describe the basic task as well as the route. All required information is passed on to the EM-platform via previously defined interfaces. Finally, the EM platform translates the requests to the AS, which executes the travel orders independently and chronologically. This is the key difference to typical driver-based DRT services: the information on the trip request is processed directly and sent to the AS (see Fig. 5). This technology is designed to be scalable, allowing larger vehicle fleets to operate in different service areas (Mook and Pavlakis 2023).

For automated PT, the tracks in Frankfurt and Karlsruhe, including the related stops, had to be incorporated as map information in the vehicle software. For smooth operation, the data must be synchronized between the vehicle software and the routing platform. Another decisive factor is the transmission of telemetry data between the platforms, for instance the speed or the state of charge of the vehicle's battery. Those data are needed for efficient vehicle planning and control. The routing technology, developed by ioki, considers the external operating conditions of the shuttles, the so-called Operational Design Domain (ODD). This includes the defined road network,

**Fig. 5** From the start, ioki's DRT software was designed to implement people-driven and automated transport systems *Source* Authors' own pictures

the maximum speed or even the minimum width for passage (Mook and Pavlakis 2023).

## 4.2 AI System for Enhancing Shuttle Services

To roll out a fully automated DRT service across a larger area, the Frankfurt site, in cooperation with T-Systems and RMV, has been testing AI systems to interact with passengers and operators. In the future, these AI systems will be integrated into the architecture of the booking and vehicle software so that all systems are smoothly connected, and passengers perceive them as a large overall system.

The AI based camera system analyses specific situations inside the shuttle and displays the output to the passengers on a monitor. The four basic use cases within SHOW were as follows: security check whether passengers are seated, left luggage, mask control (due to Covid-19 restrictions until February 2023), and passenger counting. For the future, there are many more use cases in which AI cameras can help implementing a smooth operation of fully driverless services. Additionally, as mentioned again, the voice bot was implemented with prospect to an AI-based communication tool for passengers, serving as a comfort service but also security means. In SHOW, only information about PT connections could be asked by voice inside the shuttle. Even if no additional service questions have been implemented yet, the findings on the use of AI-based voice systems were important for future implementation of similar systems.

## 5   Lessons Learned and Recommendations

In general, all activities at the German sites in Frankfurt, Monheim and Karlsruhe have run successfully and shown what sustainable and automated mobility can look like from both the aspects of operation and the vehicle development. The many positive responses from passengers indicate the success of the activities demonstrated. The spectrum of the operation duration is quite wide, from 12 months in Frankfurt and 13 months in Karlsruhe to over 3 years at the Monheim site, where in the latter, more than 200,000 kms have been travelled (considering also the pre-SHOW period). The lessons learned can be divided into several aspects: addressing citizen (and especially elderly) needs, operational matters, and the DRT service itself.

The tests of the German SHOW sites have revealed significant findings on how to address citizen needs with regards to this innovative and new form of mobility: One of the key challenges for the future success of these services will be to guide and support passengers on the path towards full acceptance of automated PT. In Monheim, the AS successfully operate as a regular bus line, which has already generated high acceptance for this new technology. Passengers can evidently integrate the automated service into their daily routine. Similarly, the service offered by the Frankfurt test site was used on a daily basis by the residents as well. Due to the demographic distribution in the area, a high percentage of the customers were elderly people. A dedicated event took place in a residential home for elderly people near the test site and helped to clarify questions and explain the new technology. The main finding was that elderly customers would be more accepting of these new technologies if communication was intense from the start, and if they felt that the local authorities and the operators acknowledge and address their worries and concerns. Overall, explaining automated technology in a simple and visual way—and especially offering the possibility to experience it—will reduce the feeling of complexity for the passengers and thus increase acceptance. Communicating with the different target groups on all three sites led to important learnings for future operation, which need to be further investigated in follow-up projects.

The operational aspect of providing AS service also needs to be considered. For simple maintenance tasks, technicians need to be trained to fix non-complex and non-software-related errors, as the handling of AS differs from that of conventional vehicles. In Monheim, it was necessary to build a garage to store and charge the AS as the bus depot was too far away from their route. The garage also protects the vehicles from theft and vandalism and shields them from extreme weather conditions which can cause battery issues. The safety operators need special training to be allowed to operate the AS, which causes extra costs and effort. Additionally, the safety operators are subject to special scheduling as they need to take more frequent breaks than drivers of conventional vehicles due to the high concentration requirements on this new field of employment. This, in turn, leads to additional costs and special scheduling when assigning safety operators. It has to be noted that such costs occur until safety operators are no longer mandatorily requested by law on board of AS. Furthermore, current technical approaches of AS rely on fixed virtual tracks, which leads to more stops of self-driving shuttles and requires manual intervention when encountering obstacles., e. g. improper parking on the roadside. Such disruption to traffic flow reduces the quality of service of the PT and may affect the acceptance of AS. Additionally, the acceptance is negatively impacted by the slow average speed, a critical factor for ensuring a reliable and efficient service.

Two out of the three German sites have been focusing on the deployment of AS as a DRT-service. Accordingly, insights were gained for the integration of an AS into existing DRT booking apps, such as the need for more flexible interaction between the different software parts. Thus, in the future, AS-software must enable the adaptation of the predefined route to react more flexibly to pooling requests. Additionally, on an operational level, some learning has been already adapted during the operations. Thus, the need for an exact positioning of the virtual stops—compared to driver-based systems—has been identified. Moreover, AS stop exactly at each predefined point, which cannot be adjusted, and a boarding/alighting problem arises if the door access is blocked by temporarily unexpected objects. Thus, the locations of virtual stops at dedicated parking spots have to be clearly analysed before the start of an automated DRT-operation. Another operational learning is that a DRT-system needs to give the related AS the task to drive to an operational stop after fulfilling a ride request if there is no follow-up mission. Otherwise, the AS remain at the assigned drop-off point, potentially obstructing traffic flow. At the Frankfurt site, this was improved by selecting virtual stops at dedicated parking spots where the shuttles could pause and wait for next ride requests.

Overall, smooth cooperation and integration of the software platforms via interfaces are essential to realise the full potential of automated mobility as part of PT and to offer passengers an efficient and comfortable DRT service. During the transient phase towards fully automated PT operation, safety operators play a crucial role in ensuring the smooth, safe and reliable operation of the shuttles and fulfilling the needs of passengers on board, as was the case at the above pilot sites. This is expected to continue until the public has a fairly good understanding of AS and AS have more robust and sophisticated technology, planning and response capabilities and on-board systems that fulfil the passengers' needs. When this is achieved, it is conceivable that remote monitoring and teleoperation could be a way to take over the tasks of safety operators. A summary of the lessons learned, and the respective recommendations is shown in Table 2.

# 6 Summary and Prospects

At the German mega site, the long-term, successfully deployed AS operations and the positive feedback from passengers and residents indicate that expanding DRT services in areas with limited access to PT is the right path to achieve a comprehensive mobility transformation. Especially for automated DRT services, it is important to involve passengers and residents, explain new service products and respond to their wishes. The pilot activities have shown that safety operators play a crucial role in the operation during the transient phase towards fully automated operation—not only due to the complex traffic situations, but also in the case of unexpected events or as a contact person for passenger requests. Continuous training and education of safety operators are important, and there is still plenty of room for improvement on the development of AS. Furthermore, the entire approval process is quite time-consuming before AS can be used in practice. There is a need to review the relevant legislation to allow easy implementation and/or integration of AS into an existing PT system. Currently, AS have a relatively low speed limit (20 km/h) in Germany. A higher speed limit can help to promote public acceptance of AS. Continuous AS dissemination and communication campaigns are also needed to achieve greater public acceptance and awareness.

Building on the knowledge and experience gained, the comprehensive implementation of automated DRT services in the Rhine-Main area will be gradually advanced and the implementation of automated driving technology in PT will be further investigated. Currently, there is no further specific AS operational plans and projects in the RMV region and at the Karlsruhe site. In Monheim, the AS are already integrated into the local PT system and will continue to offer regular services. To expand PT coverage, the provision of an additional A2 line and/or an DRT service is proposed and under discussion.

**Table 2** An overview of the lessons learned and recommendations given

| Key challenges | Recommendations |
| --- | --- |
| Smooth behaviour of AS, e.g. with less hard braking | Improvement of algorithms specifically for AS reaction mechanisms |
| Driving around obstacles | Developing AS with the ability to move freely and sophisticatedly |
| Efficient ride-pooling/dynamic routing to reduce waiting time | Adaptation of routing mid-drive for incoming on-demand requests |
| Smooth connection between the DRT software and the automated shuttles (e.g. optimizing pick-up and waiting points) | Optical marking of the exact pick-up points and programming of a fixed waiting position for the shuttle after completing a task |
| Improvement on energy consumption and driving comfort by platooning of shuttles and followers | Algorithm development for cooperative, coordinated driving manoeuvres |
| High AS maintenance effort, e.g. transporting AS back to local repair workshops | More in-house service and more support from the OEM side |
| Addressing elderly residents' needs for using booking app and reaching virtual stops | Developing and explaining complex products in a simple and visual way; Clear stops and guidance to stops; communicating with target groups to gain valuable feedback |
| Exploring new mobility concepts to satisfy passengers' needs | Prototype and product development through industry partners |
| Costly AS operation (safety operators, high maintenance costs) | Robust automated technology with high quality planning and response capabilities |
| Other road users, e.g., wrong parked vehicles as obstacles for the AS (safety stops, manual driving required) | More awareness and support from public order office: fines, less parking possibilities on routes, regularly checking on route improvement |
| Enable safety operators to use the new concept without virtual track to reduce human interventions | Continuous software-update and integrate feedbacks from safety operators |

**Acknowledgements** The work presented in this manuscript has been funded by the SHOW project which has been made possible by funding from the European Union's Horizon 2020 Research and Innovation Programme Under Grant Agreement no. 875530.

# References

Armellini MG, Banse Bueno OA, Bieker-Walz L, Erdmann J, Flötteröd Y-P, Rummel J (2021) Brunswick simulation scenario for virtual-stops based DRT services with SUMO. In: Proceedings of the 10th international congress on transportation research

Bundesministerium der Justiz (2022) Verordnung zur Genehmigung und zum Betrieb von Kraftfahrzeugen mit autonomer Fahrfunktion in festgelegten Betriebsbereichen (Autonome-Fahrzeuge-Genehmigungs-und-Betriebs-Verordnung - AFGBV) [Online]. Available: https://www.gesetze-im-internet.de/afgbv/BJNR098610022.html. Accessed 5 March 2024

Connected Automated Driving Europe (2024) Test sites [Online]. Available: https://www.connectedautomateddriving.eu/test-sites/. Accessed 5 March 2024.

DLR (2024) Test bed lower saxony for automated and connected mobility [Online]. Available: https://verkehrsforschung.dlr.de/en/projects/test-bed-lower-saxony-automated-and-connected-mobility. Accessed 5 March 2024

Dytckov S, Persson JA, Lorig F, Davidsson P (2022) Potential benefits of demand responsive transport in rural areas: a simulation study in lolland, Denmark. Sustainability 14(6):3252. https://doi.org/10.3390/su14063252

European Commission (2023) How to deal with the worker shortage and evolving skill requirements of the public transport sector—public transport and shared mobility EGUM Subgroup Topic 4B, code number E03863

Federal Highway Research Institute (2024) Digital motorway test bed [Online]. Available: https://www.bast.de/EN/Traffic_Engineering/Subjects/V5-digital-test-bed.html. Accessed 5 March 2024

Federal Ministry for Digital and Transport (2024) Digital test beds [Online]. Available: https://bmdv.bund.de/EN/Topics/Digital-Matters/Digital-Test-Beds/digital-test-beds.html. Accessed 5 March 2024

FZI (2024) EVA-Shuttle—Electric. Connected. Automated [Online]. Available: https://www.fzi.de/en/project/eva-shuttle/. Accessed 5 March 2024

FZI (2024) Shuttle2X—Safe use of automated shuttle vehicles in urban traffic through supporting infrastructure networking [Online]. Available: https://www.fzi.de/en/project/shuttle2x/. Accessed 5 March 2024

Lambing N, Och S, Orf S et al (2021) Closing the gap between automated mobility in smart cities: automated vehicle and shuttle transportation in the test area autonomous driving Baden-Württemberg," In: Proceedings of the 10th international congress on transportation research. Rhodes

Mook M, Pavlakis S (2023) EASYplus bringt die ÖPNV-Mobilität der Zukunft auf die Straße. Nahverkehrs-Praxis 9(10):30–32

Ochs S, Grimm D, Doll J-D et a (2023) Stepping ahead with electrified, connected and automated shuttles in the test area autonomous driving BW, TechRxiv. https://doi.org/10.36227/techrxiv.170327243.35964930/v1

OpenStreetMap (2024) [Online]. Available: https://www.openstreetmap.org/. Accessed 7 April 2024

SHOW (2024) SHared automation operating models for worldwide adoption: mega sites—Germany [Online]. Available: https://show-project.eu/mega-sites-germany/, Grant Agreement no. 875530. Accessed 5 March 2024

Test Area Autonomous Driving Baden-Württemberg (2024) Test area autonomous driving Baden-Württemberg—field test for research and innovation [Online]. Available: https://taf-bw.de/en/. Accessed 5 March 2024

VDV (2024) Autonome shuttle-bus-projekte in Deutschland [Online]. Available: https://www.vdv.de/liste-autonome-shuttle-bus-projekte.aspx. Accessed 5 March 2024

VRR (2024) Tariff information [Online]. Available: https://www.vrr.de/fileadmin/user_upload/pdf/service/downloads/Weitere_Broschueren_und_Tarifinformationen/VRR_Preise_2024.pdf. Accessed 5 March 2024

# Autonomous Bus Depot Management: Operator's Lessons Learned and Cost Analysis Perspective

César Omar Chacón Fernández, Sergio Fernández Balaguer, Lucía Isasi de la Iglesia, and Borja Gorriz Espinar

**Abstract** This study delves into autonomous bus depot management at EMT Madrid, the Madrid Public Transport Company, within the Horizon 2020 SHOW project (GA No 875530). Its aim is to boost operational efficiency and optimize resource usage, particularly by addressing unproductive hours for bus drivers. It examines the technical feasibility of automating manual depot tasks like vehicle charging, cleaning, and parking through advanced sensor technology and control systems. The pilot at Carabanchel bus depot involved automated vehicles (AVs) with perception sensors, control mechanisms, and centralized decision-making units, testing services like internal transport, autoparking, and teleoperation. Over eleven months, the AV fleet performed successfully without incidents. The findings indicate automation's promises in reducing operational costs and enhancing resource utilization, though, challenges like initial investments, technical constraints, and regulations persist. Recommendations are made to foster public–private collaboration for innovation in public transport and for market development.

C. O. C. Fernández
Head of Engineering projects Department. Transport Services Directorate, Empresa Municipal de Transportes de Madrid S.A. (EMT Madrid), C/Ventura Díez Bernardo S/N, 28054 Madrid, Spain
e-mail: cesaromar.chacon@emtmadrid.es

S. F. Balaguer (✉)
Head of International and EU Projects Department, Institutional Affairs and ESG Directorate, Empresa Municipal de Transportes de Madrid S.A. (EMT Madrid), C/Cerro de La Plata 4, 28007 Madrid, Spain
e-mail: sergio.fernandez@emtmadrid.es

L. I. de la Iglesia
Fundación Tecnalia Research & Innovation, Industry and Mobility, TECNALIA, Astondo Bidea, Edificio 700, 48160 Derio (Vizcaya), Spain
e-mail: lucia.isasi@tecnalia.com

B. G. Espinar
Project Engineering and Innovation, Systems Engineering Department, Irizar E-Mobility (IRIZAR), IRIZAR, Erribera Industria Gunea, 1, 20150 Aduna, Gipuzkoa, Spain
e-mail: bgorriz@irizar-emobility.com

© The Author(s) 2025
H. Cornet and M. Gkemou (eds.), *Shared Mobility Revolution*, Lecture Notes in Mobility, https://doi.org/10.1007/978-3-031-71793-2_6

**Keywords** Autonomous bus depot operations · Connected cooperative and automated mobility · Remote operation · Cost efficiency · Operation optimization

# 1 Introduction

Empresa Municipal de Transportes de Madrid S.A. (hereinafter EMT Madrid) is a public company that manages the entire bus transport service in the municipality of Madrid in Spain. In 2024, its fleet accounts 2,100 buses with 4,097.68 km network and since December 2022 all the fleet is 100% green (either electric or using compressed natural gas) according to the European Clean Vehicles Directive (European Parliament 2019).

Automated Mobility is one of the fields in which EMT Madrid wants to move forward, being even explicitly stated within its own Strategic Plan 2021–2025 (Madrid and Estratégico 2021).

For Public Transport Operators (PTOs), automation presents a significant opportunity to enhance the quality of service for passengers by ensuring more frequent and reliable transport options. It also promises to elevate operational safety through advanced sensor technologies and to optimize operational expenses. This latter point is especially crucial for public entities, which commonly face financial constraints. Central to the discourse on cost optimization is an unexpected focus: the bus drivers. While many cities report difficulties in recruiting bus drivers, citing a widespread driver shortage (UITP 2023), this issue is, so far, notably absent in Madrid. Instead, inefficiencies have been identified in the management of bus depots, where drivers often spend excessive amounts of time waiting for the vehicle to charge or to be cleaned, rather than actively transporting passengers. EMT Madrid's analysis reveals that each driver spends approximately 30 min/day in such unproductive task slots. On average, there are around 350 bus drivers/per bus depot, each one assuming a 22 day working month, this results in a total of 3,337 unutilized hours each month. This chapter explores the potential to reclaim these hours of inactivity through the automation of bus depot operations. It outlines the necessary technological upgrades for vehicles (Sect. 4), shares operational insights gained from the trials conducted within SHOW[1] (Sect. 5), and conducts a comprehensive analysis of the financial implications of these changes (Sect. 6).

---

[1] https://show-project.eu/.

## 2 Automated Bus Depot Management Use Case at EMT Madrid

Within the Madrid Mega Site of SHOW, the local consortium was formed by EMT Madrid and the cooperation of TECNALIA (technology provider and coordinator of Madrid Mega Site), INDRA (telecommunication provider) and IRIZAR (bus manufacturer, provider of a 12 m bus).

Although the technology on autonomous vehicles is constantly boiling and is a highly topical issue, from a scientific-technical point of view, operation with a certain degree of automation in mid and regular buses (8 m or longer) is still very limited, in addition to the lack of offer of autonomous buses in the market, especially not in 2020 at the beginning of the SHOW project. Taking into account the aforementioned limitations, the autonomous bus depot management use case has focused on process automation to achieve more efficient resource management within the bus depots, with numerous bus driver hours considered as non-productive yet with a high potential for increased efficiency.

### 2.1 Objectives of the Autonomous Bus Depot Piloting

The main objective has been to evaluate the technical feasibility of automating operations that are currently carried out manually, to improve the efficiency and to optimize the functioning of the bus depot in many ways, such as reducing the space used for parking, reducing the number of non-productive hours of bus drivers driving inside the bus depot upon arrival at the end of its service, and improving road safety.

The autonomous operation is broken down into the following tasks:

- Navigating in autonomous mode between the main point of interest of the bus depot: The washing area, the workshop (if maintenance is needed), the parking area for charging and until the bus's next service, the main entrance
- Providing an internal mean of transport for employees and occasional visitors
- Monitoring (and remote driving in specific situations) during autonomous operations

### 2.2 Required Milestones

The technical objective has been, in short, to provide the autonomous vehicles with the physical and logical means to be able to safely automate the aforementioned operations.

This requires the following development and integration steps:

1. Retrofit of the bus with perception sensors (camera, LIDAR), positioning equipment (GPS, Inertial Measurement Unit-IMU), performance data collection unit (vehicle speed, steering angle…), and integrate V2I (Vehicle to Infrastructure) communications devices, developing the correspondent decision-making logic system.
2. Automate bus controls, i.e., accelerator, brake and bus steering, as well as starting and parking brake.
3. Develop high-definition digital map of the bus depot.
4. Develop a location and navigation system in open and semi-covered areas.
5. Define and record exhaustive datasets to feed the automatic learning system that allows the vehicle to solve all the usual situations in the operating environment.
6. Develop a vehicle management station to make the appropriate calls to each of the operations (refuelling, cleaning, workshop, parking, etc.), as well as to resolve a specific circumstance that the automatic control system does not deal with or that the operator considers unsafe.
7. Integrate all the elements developed and validate them in EMT facilities.

## 3   The Carabanchel Bus Depot

Carabanchel bus depot is one of the five EMT bus depots, internally called "Operations Centres" (Fuencarral, La Elipa, Sanchinarro, Entrevías and Carabanchel), with a surface area of 65,000 $m^2$ and housing 458 buses serving 48 bus lines, including CNG and fully electric bus units. It is also where most of the electric bus fleet of EMT is based. It is located in the south of the city and is currently a benchmark in terms electrification and smart charging (170 inverted pantographs, divided into two facilities/phases: Phase II with 52 inverted pantographs and Phase III with 118 ones), while Phase IV (and last one to reach full electrification of the bus depot) is under construction at the moment of writing this article.

This bus depot offers:

- Semi controlled area with interaction with other non-autonomous buses and vehicles, as well as daily operations at the depot (manoeuvring, moving goods, people, etc.).
- As for SHOW testing track, it includes a round trip of 800 m connecting different facilities within the bus depot, so called "5 stops service".
- The depot traffic environment is equivalent to an urban one, with interaction with car/bus/trucks and pedestrians.

The operation area is outside the built-up one, with a predominant bus traffic composition. However, there is also a mix of cars (service cars) and trucks, as well as pedestrians (employees walking around by foot) interacting under different traffic conditions, as the traffic density varies across the area and across the day, considering depot rush hours very early in the morning and evenings. There is not an internal traffic control system within the bus depot. There are some vehicle movers that drive buses inside the depot from one point to the other specific place (e.g., parking slot to maintenance area).

The road type is equivalent to an urban one, with one or two lanes (depending on the area), and 5 intersections. The speed limit within the bus depot is 10 km/h, dropping to 5 km/h in the bus cleaning area, and 20 km/h on the bus testing circuit. During the whole duration of the project there has been ongoing road works in the upper terrace for the building of the different phases of the charging infrastructure for e-buses (as Carabanchel bus depot is under a full electrification process scheduled to be completed by the end of 2024, or early 2025).

In general, the sight conditions are normally clear, with some glaring depending on time-of-day. Traffic (both coming from vehicles and pedestrians) may be hidden by parked buses. Regarding the weather, in Madrid is mostly dry and sunny.

## 4 Setting the Autonomous Fleet

### 4.1 Retrofitting the Bus Fleet with Autonomous Features

Carabanchel scenario has operated three AVs: two Tecnobus Gulliver microbuses provided by EMT Madrid and one i2eBus provided by IRIZAR. TECNALIA has also provided two Renault Twizy though they have been used only during development phase, to transfer the verified and technically validated AV algorithms to the minibuses (Gulliver) and the 12m bus (i2eBus). Figure 1 depicts the three types of AVs at Carabanchel scenario with different SAE[2] and TRL[3] levels, maximum passengers' capacity, enhanced technologies upgrades and with trials' speeds.

Two of the vehicles (one Gulliver minibus and the IRIZAR bus) have been retrofitted with sensor perception, bus control mechanisms (mechatronics), and a central decision-making unit. The second Gulliver minibus has been retrofitted in a slightly different way, to also allow teleoperation (remote controlling).

***Tecnobus Gulliver minibuses***

*Localization and perception sensors*

- 4 LIDAR sensors (two front and two rear) for primary obstacle detection.
- 2 side-view cameras for monitoring surrounding areas of the vehicle and merging information with primary sensors.
- 2 video cameras (one front and one rear) transmitting real-time images to the control station for evaluating potential unexpected situations or for teleoperation.

---

[2] Refers to the Society of Automative Engineers (SAE) J3016 standard, that defines six levels of automation, from 0 to 5, source: 2024 SAE International (2021).

[3] Technology Readiness Levels (TRL) are a type of measurement system used to assess the maturity level of a particular technology (Commission and "Technology Readiness Levels (TRL)" 2024).

Fig. 1 AVs fleet used in Carabanchel bus depo. *Source* Authors' picture

- Vehicle positioning system via GPS and inertial sensors (although this positioning can be refined through visual odometry).
- Control Mechanisms (or low level control): The automation system intervenes in steering, speed control and gear selection. It utilizes redundant control mechanisms to ensure safety and reliability, such as duplicated pedal signals for speed control and hydraulic pumps for brake simulation. An electric motor assists in steering, with a clutch mechanism for manual override.
- Central Decision-Making: A programmable logic controller (Programmable Logic Controller (PLC), which is a small, modular, solid-state computer with customized instructions for performing a particular task) serves as the central decision-making unit, processing sensor data and issuing commands to robotic elements. Military-grade connectors ensure secure and watertight connections.
- Human–Machine Interface (HMI): To ensure the driver has awareness and control of the autonomous operations, an HMI is installed with all relevant information and commands.
- Operational Data Collection: The system collects operational data such as speed and steering angle for analysis and monitoring purposes.
- Communication Systems: Dedicated communication systems enable command transmission and real-time video streaming for teleoperation (Fig. 2).

### *IRIZAR i2e (I2EBAR) bus*

*Localization and perception sensors*

- 3 LIDAR sensors for primary obstacle detection.
- 3 Radar sensors for secondary obstacle detection. Normally not in use due to the amount of data to be transmitted.

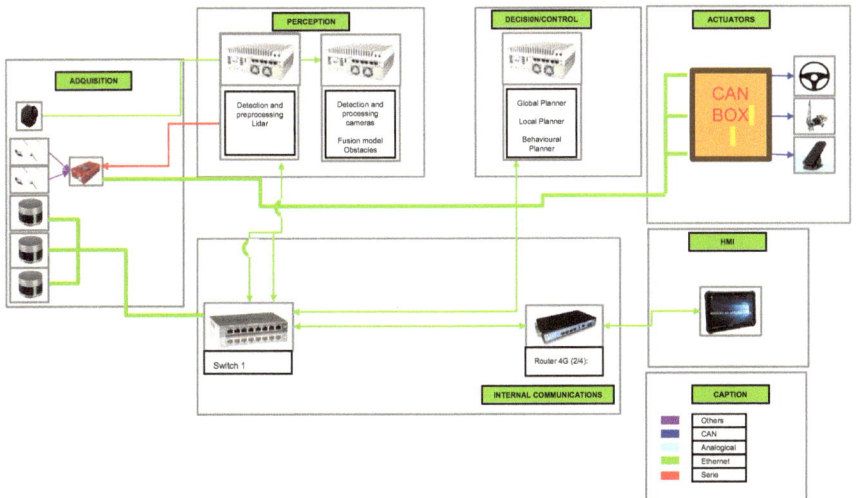

**Fig. 2** Decision and control schematic for the Gulliver minibus. *Source* Authors' picture

- 2 side-view cameras for monitoring surrounding areas of the vehicle and merging information with primary sensors.
- 1 frontal camera for obstacle identification.
- Vehicle positioning system via GPS and inertial sensors (although this positioning can be refined through visual odometry) (Fig. 3).
- Control Mechanisms: The automation system intervenes in steering and speed control. In the case of the Irizar vehicle, the gear selection is still manual. Handover strategies are in place to ensure the driver has control over the autonomous operation at any moment. The longitudinal control (in terms of acceleration) of the vehicle is done via CAN. The Control ECU intercedes in the CAN Bus and modifies the command done by the driver while in autonomous mode (Fig. 4).

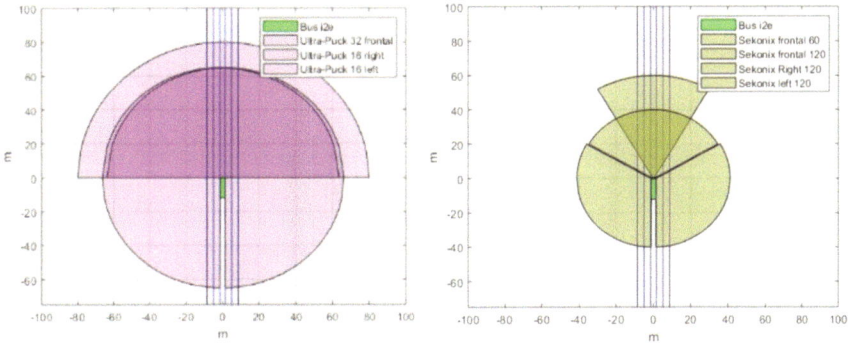

**Fig. 3** Overview of the LIDAR (Left) and Camera (Right) coverage for the I2EBAR bus

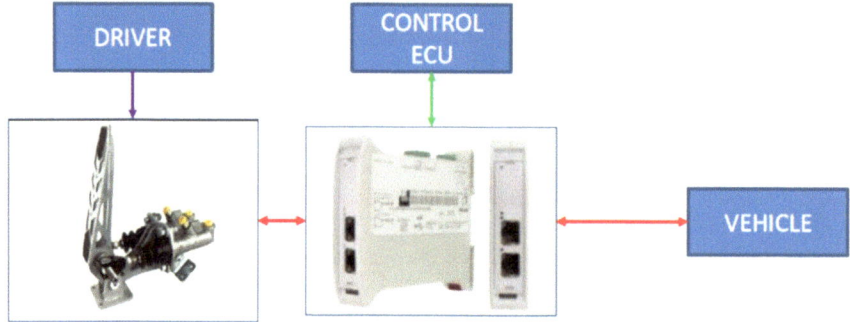

**Fig. 4** Throttle pedal control schematic for the I2EBAR bus. *Source* Authors' picture

   The longitudinal control (in terms of braking) of the vehicle is done physically. The Control ECU control a DC motor which then actuates on the brake pedal (Fig. 5).
   The lateral control is done by actuating physically on the steering shaft. The actuator consists on a DC motor, an electromagnet that acts as a clutch and a torque sensor (for the handover strategies). While in autonomous mode, the electromagnet is engaged, which means that the motor can actuate on the steering shaft (Fig. 6).

- Central Decision-Making: A programmable logic controller (PLC)/industrial computer serves as the central decision-making unit, processing sensor data and issuing commands to robotic elements. Military-grade connectors ensure secure and watertight connections (Fig. 7).
- Human–Machine Interface, HMI: To ensure the driver has awareness and control of the autonomous operations, an HMI is installed with all relevant information and commands.
- Operational Data Collection: The system collects operational data such as speed and steering angle for analysis and monitoring purposes.

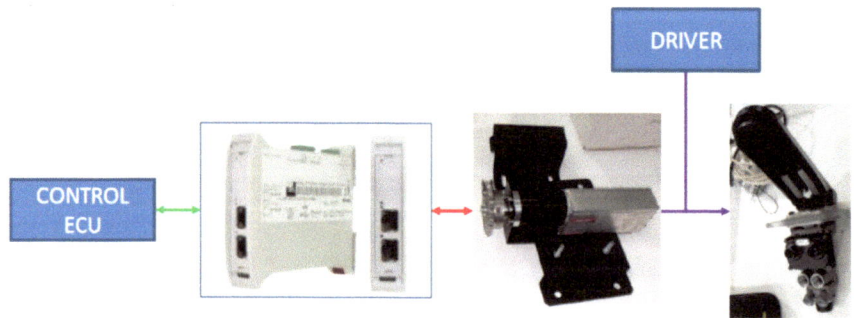

**Fig. 5** Brake pedal control schematic for the I2EBAR bus. *Source* Authors' picture

**Fig. 6** Lateral control (in vehicle) for the I2EBAR bus. *Source* Authors' picture

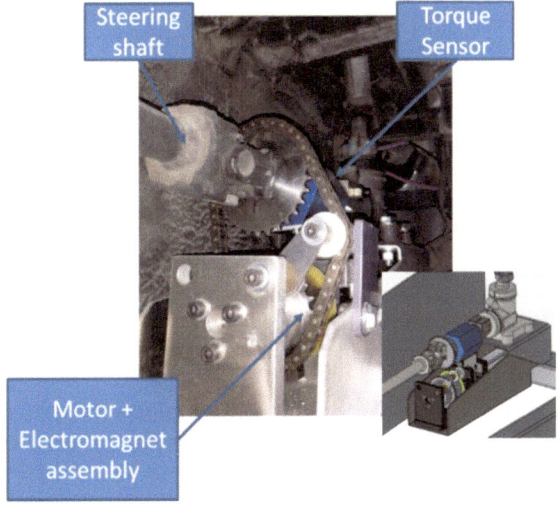

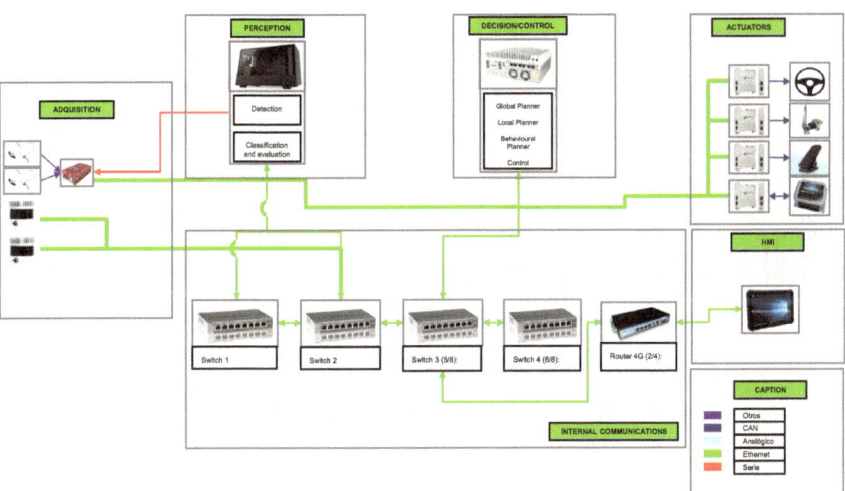

**Fig. 7** Decision and control schematic for the I2EBAR bus. *Source* Authors' picture

As a conclusion, for both types of buses, the Tecnobus Gulliver and the Irizar bus, the focus has been the integration of advanced technologies to achieve safe and efficient autonomous operation. By leveraging sensor perception, robust control mechanisms, and centralized decision-making, the system promises to revolutionize public transportation systems with enhanced safety and reliability. However, this process is both complex and lengthy.

# 5   Services and Vehicle Performance

Within the Carabanchel bus depot use case, Madrid SHOW partners have deployed, tested and validated three different services:

- "*5-stop service*": by using one Gulliver minibus and the I2ebus provided by Irizar, an internal and regular mean of transport has been provided for EMT Madrid employees and visitors.
- Autoparking service: parking buses without the intervention of a bus driver, in an autonomous way.
- Teleoperation service (both supervision and actuation on the vehicle): driving remotely one Gulliver minibus from a control centre situated in the main building.

It is important to remark that training sessions have been a priority to all personnel involved before launching any of the services.

## 5.1   Five Stops Service

As indicated previously, the "*5 stops service*" has been deployed to provide transportation within Carabanchel bus depot, connecting the 5 points of interest of the bus depot where most of the employees flow concentrates within the facility. The "*5 stops service*" route consists of an 800 m long round trip (circular), giving public transport service to EMT employees as well as depot external visitors. Key attention points regarding this are as follows:

- There is always some level of traffic going inside the depot, including depot service non-autonomous vehicles, employees' vehicles and employees moving as pedestrians. The employed AVs fleet should be able to traverse this scenario in autonomous driving mode without safety incidents. In specific, the "*5 stops service*" consists of a round trip which includes pedestrian crosses, which have to be successfully negotiated by the AVs.
- The operational part of the scenario contains some complexity as the service is expected to operate during rush hours at the depot, when there is increased traffic density, given that most routes start and end at similar hours, making bus traffic much higher. AVs are expected to provide the service without safety incidents, and with minor delays or increased stops due to the traffic.
- This service has successfully run in weekdays making use of one Gulliver minibus and the 12m i2eBus. The experimental process was as following:
- Early in the morning, considered rush hour inside the depot, both Gulliver and i2eBus safety drivers drive manually their AV outside boxes (overnight parking lots), towards the first stop, i.e. Main Building. Once at the first stop, the drivers start the service by using their dedicated HMI, changing to autonomous mode.

- The "5 *stops service*" collects/drops EMT employees at any of the 5 stops, on a round trip basis. While driving around the depot, always at allowed speeds, it overcomes non-autonomous vehicles as well as pedestrians.
- As soon as a passenger gets on the bus, the safety driver takes this into account (occupancy); while when the passenger gets off the bus, when applicable, denotes on the dedicated HMI the satisfaction (against a specific 5 levels scale).
- The service runs until the afternoon. Then, the Gulliver bus is manually driven back to boxes for charging/maintenance. While i2eBus is parked automatically in its designated parking spot (see Sect. 5.2 autoparking service), until late afternoon, when is manually driven back to boxes for charging.
- Wednesdays is a particular day at the depot as it is students' day visit. SHOW has taken advantage of this (almost) weekly event to present the project to young generations as well as a tour around the depot, visiting the different areas (training, boxes, main building, etc.) (Fig. 8).

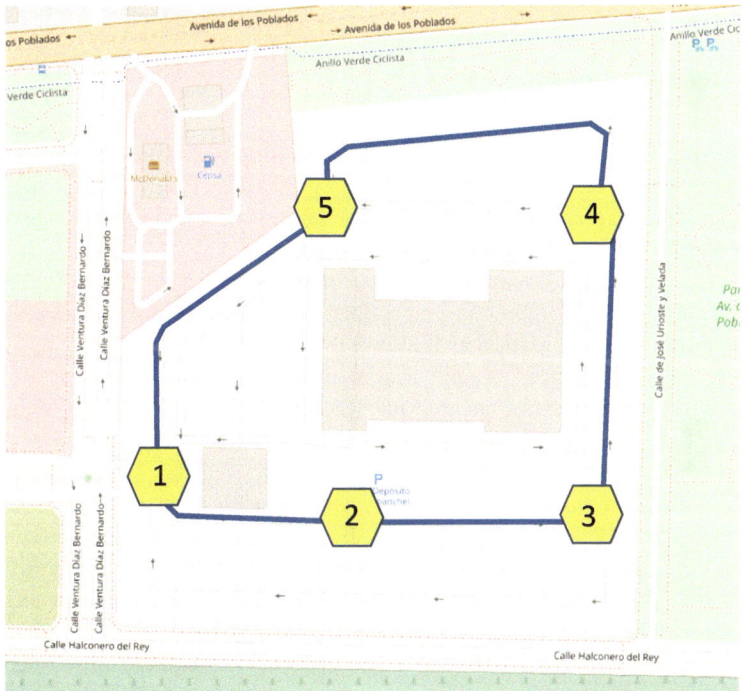

**Fig. 8** Five stops route (800 m): (1) Main bus depot entrance, (2) Main building personnel entrance, (3) Upper terrace bus parking and charging area, (4) Secondary bus depot entrance, (5) Training area. *Source* Open Street Maps

## 5.2   Autoparking

The second service provided is the autoparking function. This service has been operated by the 12 m i2eBus IRIZAR bus. This use case is meant primarily to manage the depot and autopark the e-bus during the day, once the 5 stops service is completed (sometimes in the afternoon). For this purpose, i2eBus' safety driver selects in its dedicated HMI the designated parking spot out of the three allocated for this service (parking lots 916, 917 and 918, all located in the lower terrace of the bus depot). Next, the 12 m bus is able to drive in autonomous mode to the selected slot, prior to performing the parking manoeuvre; and next to park in the selected slot automatically (no driver's intervention). Worth mentioning that at the end of their shift, if applicable, it has been considered that safety drivers report any incident that occurred (offline, writing on a notebook), though no significant one occurred during the piloting.

## 5.3   Teleoperation

This service has been tested with one Gulliver minibus which has been instrumented with four cameras (AV's environment perception) and Wi-Fi access (able to communicate with the remote-control desk). Teleoperation service includes both supervision and actuation on the vehicle.

The remote driving operation is done from a remote desk, where the technical personnel in charge is remotely controlling the Gulliver. This functionality is of great interest for EMT Madrid as it allows the possibility of managing buses at the bus depot without the need of a physical interaction with the vehicle.

From the communication perspective, EMT has based this use case in the Wi-Fi coverage that the Carabanchel bus depot has.

The control desk has been installed in the main building of Carabanchel bus depot, in one of the staff offices, from which it is possible to take control or supervise autonomous or manual driving.

This control desk could be potentially integrated into the central control centre (Servicio de Ayuda a la Explotación, S.A.E.) that EMT has at its main headquarters, from which all the on-street bus operations are managed.

To carry out this task, the following equipment was required to set the control desk, screens, computer for sending missions and Wi-Fi connections and connectors (Fig. 9).

Teleoperation requires preliminary studies, vehicle adaptations, and verification tests. Initially, communication methods were explored, with Wi-Fi technology being the most effective. A specific network band was allocated after assessing available options. An algorithm ensures the bus connects to the access point with the strongest signal, reducing delays. Latency was measured for optimization. The same infrastructure for vehicle movement is used for autonomous and teleoperated driving. The

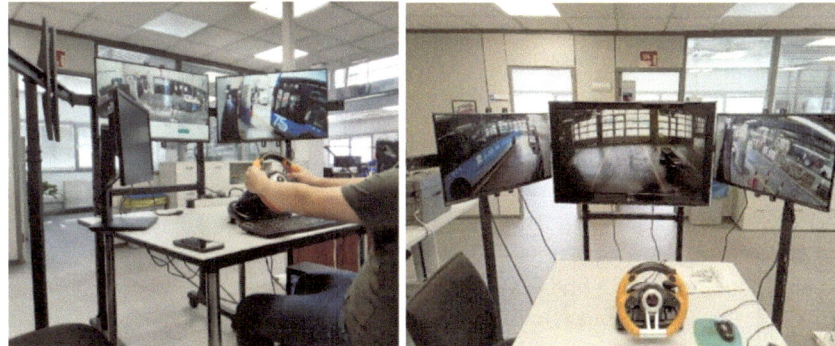

**Fig. 9** Teleoperation control desk in Carabanchel bus depot. *Source* Authors' picture

control desk includes screens for real-time vehicle images, a high-performance PC, and a steering wheel for a realistic driving experience. It allows for switching between autonomous and teleoperated modes.

## 6 Fleet Performance and Cost Analysis

### 6.1 Fleet Performance

Regarding the fleet performance, all the services, testing and piloting have run smoothly and without any incident during the 11 months of operation, with a total number of 6,363 passengers transported and a total of 12 different safety drivers.

Concerning the fleet characteristics, as a summary, the minibus Gulliver has SAE level 3–4. Enhanced perception and decision driving functions. The maximum speed is 10 km/h with 7.2 km/h on average. The passenger's capacity is 11 (safety driver included). Irizar bus has SAE level 4 with the same upgrades but a higher speed limit (18 km/ maximum and 8.96 km/h average). Its capacity is 26 seated people, safety driver included.

### 6.2 Cost Analysis

In the context of SHOW, a preliminary cost analysis has considered both costs and savings assuming the automation of the bus fleet at a generic bus depot such as the Carabanchel one, exclusively considering the autonomous operations within the bus depot. It is important to remark that this is a rough initial calculation which, potentially, could be much more elaborated.

- Bus drivers:

According to the collective agreement with the human resources department, the trade unions and the operational requirements established in the company, bus drivers have up to 13 min to enter and 13 min to leave the bus depot upon arrival at the end of their journey (this is the official time they have allocated within their labour journey). That is a total of 26 min/day per driver. On average, there are 350 drivers per bus depot, and 22 working days in a month. That makes a total of 9,120 min/day (152 h/day), 200,220 min/month (3,337 h/month) that potentially could be saved if operations become autonomous. To translate that into euros, and considering an average salary of a bus driver with 14 years in the company, the salary is 43,178.47 €/year without considering social security, and 57,155.34 €/year considering social security; this second salary including social security is the cost for the company. Considering an average of 22 working days a month and an average of 8 h a day, drivers work a total of 2,112 h/year. If we divide the salary by the hours, this results in an average salary cost of 27.06 €/h. Going back to the saved hours and multiplying them by the hourly cost, automation could save up 4,113.2 €/day per bus depot for the specific role of bus driver waiting at the depot (90,488.64 €/month for a bus depot like Carabanchel). Safety drivers' costs are not included as, within the bus depot, they would not be necessary.

- Remote operators (teleoperation):

In the event that the autonomous bus encounters a situation where it cannot act, a remote operator would take control. Remote operations could be performed by the current "vehicle-movers" staff. This specific category of staff takes care of moving buses around the depot, for whichever reason needed. That makes 6 persons per day that should be trained to upgrade their skills in order to become remote operators. These workers have an average salary 20% lower than that of the bus drivers, so we can consider a cost for the company of 45,724.27 €/year per each. Considering those 6 persons, remote operators would mean a total cost of 274,345.62 €/year. However, this is already existing personnel, and therefore this cost is not included within the savings calculation. Simply, vehicle movers will be turned into remote operators after the proper training.

- Engineering costs:

Assuming that the retrofitting of buses would be done "in-house", the company assumes an average dedication of 950 h/year dedicated to define, plan and execute the automation tasks from our engineers. Engineers have an average salary of 27.06 €/h, meaning a cost of 25,707 €/year for the engineering process. This task and cost would affect only at the beginning of the retrofitting project (just one time) as this initial development would serve as the foundation for future developments on other bus models.

- Mechanic workers:

The mechanics are responsible for assembling the components on the bus and also for repairing them when breaking down or misaligning. A similar average salary as for the "vehicle movers" is considered, as well as an average dedication of 40% of their time for these tasks. It should also be noted that dedication hours are higher at the beginning of the process (learning curve). Therefore, for the "in house" automation of vehicles there would be an average cost of 18,289.71 €/year per mechanic worker.

- Equipment:

For the current calculation, the investment on equipment for retrofitting one single bus is estimated to be approximately 75.000 €/vehicle. This cost includes the purchase of the equipment indicated in Sect. 4.1, but not the hypothetical maintenance of it.

- Total cost:

Despite the figures, the calculation of the total cost analysis is not evident nor easy to obtain, as it is important to consider the following limitations and considerations:

– Occupations of workers: the 26 min "saved" at the depot cannot be transferred that easily to another occupation. "Rigidity" of the contracts and laws, made this option not that flexible nor possible. However, one of the purposes of this paper is to highlight the potential for more efficiency in bus depot operations by automation.
– Price of technology is not entirely predictable. It is envisioned that technology will become cheaper as soon as more manufacturers engage. However, this has not happened yet.
– The initial upfront cost of purchasing the equipment for 450 buses (average amount of buses in one bus depot) is considerably high and makes the calculation quite complex as it should consider the correspondent depreciation. With the 75,000 € considered per bus, the total cost to retrofit the whole fleet at Carabanchel bus depot would sum up to 33,75 million euro plus mechanic and engineering costs (roughly estimated in 8.2 additional million €).
– It is important to consider that in the Madrid bus depot pilot, vehicles have been retrofitted by the partners involved. The costs reflect the hurdles and challenges of retrofitting a vehicle for the first time, and in this case, it should also be taken into account that the retrofitted vehicles were already quite old and therefore giving extra issues which demanded more man-hours than expected.
– In the future, EMT Madrid (as a public transport operator) is considering on directly acquiring the autonomous vehicles from the manufacturer. According to Ongel et al. (2019), it is suggested that the acquisition prices are estimated to be 20% higher for buses with autonomous functions in 2030.

Therefore, despite the evident savings in personnel costs (up to 1 million € per year per bus depot), it is not possible to conclude with a proper cost analysis of costs and savings. The high upfront cost of purchasing the equipment and retrofitting conventional buses, shows a low cost/benefit ratio for adapting existing non autonomous buses (around 26% of the initial cost of an e-bus plus the cost of the e-bus itself), and the uncertainty about some aspects, including depreciation, makes extremely complex to provide concluding figures.

## 7 Conclusions and Lessons Learned

The focus of the Carabanchel pilot in SHOW was on the exploration of the potential optimization of bus depot operations while at the same time making AVs more attractive to general public by deploying public transport services.

The main impact is the learning on how automation can help improving efficiency in bus depot management, and the confirmation that technology is mature enough to be deployed in semi-controlled environments such as a bus depot, bringing also further possibilities with additional operations such as charging processes.

Public private cooperation is key in this process, and operators, manufacturers and other stakeholders involved need to maintain it to help the market provide more (and cheaper) autonomous solutions. Regarding the technology, it is evolving quickly and new manufacturers will enter into the market enabling prices to go down and making this kind of operations accessible on a commercial basis.

Finally, several the lessons learned and recommendations are proposed for replicability. The cost–benefit ratio for adapting existing non-autonomous buses is relatively low, approximately 26% of the initial cost of an e-bus plus the cost of the e-bus itself. There is significant potential for scaling up, making it worthwhile to initiate discussions with bus manufacturers to test further solutions. Public–private cooperation is essential, and exploring the potential addition of autonomous charging processes using inverted pantographs could be beneficial. General acceptance, particularly among bus drivers, was wider than expected, likely due to the perception that broader automation is still in the future. In this context, it is crucial to plan training sessions for labour forces well in advance.

From the perspective of bus manufacturers, there is potential for new business lines. Efforts should focus on industrialization and deployment, with further research and innovation projects needed. The market needs to develop further to provide competitive products and enable cost-effective business models.

**Acknowledgements** The work presented in this manuscript has been funded by the SHOW project which has been made possible by funding from the European Union's Horizon 2020 Research and Innovation Programme Under Grant Agreement no. 875530.

## References

European Commission "Technology Readiness Levels (TRL) (2024) Accessed 26 April 2024. https://ec.europa.eu/research/participants/data/ref/h2020/other/wp/2016_2017/annexes/h2020-wp1617-annex-g-trl_en.pdf

European Parliament (2019) Directive (EU) 2019/1161 of the European Parliament and of the Council of 20 June 2019 on the promotion of clean and energy-efficient road transport vehicles. Available at https://eur-lex.europa.eu/eli/dir/2019/1161/oj. Accessed 19 March 2024

EMT Madrid (2021) Plan Estratégico 2021–2025 (In Spanish). Available at https://www.emtmadrid.es/Ficheros/Portal-Transparencia-2021-B/Plan-Estrategico-de-la-Empresa-Municipal-de-Transp.aspx. Accessed 19 March 2024

Ongel A, Loewer E, Roemer F, Sethuraman G, Chang F, Lienkamp M (2019) Economic assessment of autonomous electric microtransit vehicles. Sustainability 11(3):648

SAE International (2021) J3016_202104: ground vehicle standard taxonomy and definitions for terms related to driving automation systems for on-road motor vehicles. https://doi.org/10.4271/J3016_202104. Issuing Committee: On-Road Automated Driving (ORAD) Committee. SAE International, Publisher, pp 41

UITP (2023) Business and human resources news. Accessed 1 Dec 2023 from https://www.uitp.org/news/new-uitp-taskforce-on-workforce-shortage-tackles-transforming-labour-market/

# Field-Driven Lessons Learned for Delivery Robots Logistics Operation in Urban Environment

**Venkata Akhil Babu Malisetty, Elena Patatouka, Odisseas Raptis, Anna Antonakopoulou, Angelos Amditis, Enrico Silani, and Elvezia Maria Cepolina**

**Abstract** In the city centre of Trikala, Greece, a real-life logistics service was operated with a fleet of 5 droids on a public route in the urban pedestrian area in the context of the SHOW project (GA No 875530). It took place from December 2022 to the end of February 2023. The droids are small automated vehicles missioned to carry out indoor and outdoor logistic services in pedestrian-centric environments. The logistics service that operated in Trikala was of two types: (a) a multi-stop parcel delivery service from the depot to the shops and (b) a parcel collection service from the shops to the depot. The current manuscript describes the solutions implemented and the challenges faced by the droids while performing last-mile deliveries. The manuscript concludes with suggestions from experience for new implementations of logistics services operated by automated vehicles. These include the advice to not underestimate the difficulty of integrating radically new vehicles into existing

V. A. B. Malisetty
DIME, University of Genova, Via Opera Pia 15, 16145 Genoa, Italy
e-mail: venkataakhilbabu.malisetty@edu.unige.it

E. Patatouka · O. Raptis
e-Trikala S.A, Kalampakas 28 and Ampatis, 42100 Trikala, Greece
e-mail: elpatatouka@e-trikala.gr

O. Raptis
e-mail: oraptis@e-trikala.gr

A. Antonakopoulou · A. Amditis
National Technical University of Athens, ICCS, 9, Ir. Politechniou Str., 15773 Zografou, Athens, Greece
e-mail: anna.antonakopoulou@iccs.gr

A. Amditis
e-mail: a.amitis@iccs.gr

E. Silani
e-novia, Yape S.R.L., Via San Martino 12, 20122 Milan, Italy
e-mail: enrico.silani@e-novia.it

E. M. Cepolina (✉)
Elvezia Maria Cepolina, DIME, University of Genova, Via Opera Pia 15, 16145 Genoa, Italy
e-mail: elvezia.maria.cepolina@unige.it

© The Author(s) 2025
H. Cornet and M. Gkemou (eds.), *Shared Mobility Revolution*, Lecture Notes in Mobility, https://doi.org/10.1007/978-3-031-71793-2_7

logistics services; the importance of having the trust and support of the population and the need for data network stability.

**Keywords** Automated logistics · Small automated vehicles · Last mile deliveries · Real life operations

# 1  Introduction

Due to rising expenses for urban storage areas and the growth of the global e-commerce industry, there is a greater evolving need for parcel delivery services. Last mile delivery in populated urban areas is experiencing huge challenges because of worsening traffic congestion, limited parking spaces, and environmental constraints (Akeb et al. 2018).

Last-mile delivery has been a crucial success element for Logistics Service Providers in achieving high customer satisfaction and growing market share (Cepolina et al. 2021).

New last mile delivery strategies have consequently been developed over the past ten (10) years. These new solutions combine electric propulsion with automated driving. Up to 80% of all Business-to-Customers (B2C) deliveries are anticipated to be amenable to automated products delivery. Deliveries with automated vehicles (AVs) begin to be made both by air and by land (Grolms 2019; Chen et al. 2021).

Unmanned Aerial Vehicles (UAVs) are already being employed in actual tests by trailblazing businesses including DHL (Heutger and Kückelhaus 2016), SF Express (Shields) Amazon (Vincent and Gartenberg 2019), Google (BBCNEWS 2018) and UPS (McFarland 2019). However, UAVs are unable to do deliveries in some locations due to their low carrying capacity, limited flying range, and legal constraints in metropolitan areas.

Referring to AV, Reed et al. (2021) model an automated vehicle that can drop off the delivery person at selected points in the city where the delivery person makes deliveries to the final addresses on foot. Then, the vehicle picks up the delivery person and travels to the next reloading point. In this way, the delivery person would never need to look for parking or walk back to a parking place. Another proposed solution for last-mile delivery service, is the AV with lockers, which can deliver packages without human assistance. Noticeable services are the ones operated by Nuro delivery vehicle (Hawkins 2024) and by Starship Technologies. Starship Technologies is now operating a delivery robot service in the Northern Arizona University campus (Gehrke et al. 2023). Customers are informed of the precise arrival time and instructed to retrieve a package from a designated locker mounted on an AV (Bouton et al. 2017). A similar experimental system is proposed by Cepolina (2016) where the box is left for a time window in a bay chosen from time to time according to the addresses of the parcels in it. This box is transported into the bay and here autonomously unloaded by the FURBOT prototype (Molfino et al. 2015), a compact and environmentally friendly vehicle developed in the FURBOT FP7 European project. The land occupation is minimized because the unloading bay is only occupied by the box when there is freight addressed to it; otherwise, it can be used for other purposes, such as parking. The box's high loading factor reduces the number of trips required for urban freight distribution, which will have a positive effect on pollution. Other delivery services operated by AV are those involving Yapes: in Stockholm they carried out home food deliveries accompanied by a human co-pilot and at Frankfurt airport (ROBOTICS and AI 2019) indoor delivery services. Other experiments with Yape were con-ducted in Japan, in the Fukushima district (Deragni 2019) and in Italy, in the UpTown Smart District in Milan (Altea 2022).

In the context of the SHOW project (SHOW European project), logistics services operated by AVs have been planned and implemented in Carinthia (Austrian state), Trikala (Greece), Karlsruhe (Germany), Gothenburg (Sweden) and Geneva (Switzerland).

This paper focuses specifically on the logistic services operated in the Greek city of Trikala. The pilot carried out in Trikala City in SHOW has been very challenging in relation to the context (public ground, in the city centre, without the constant presence of a co-pilot), in terms of duration (services operated for 3 months) and in relation to the size of the fleet involved (a prototypal fleet of 5 small AVs). The experimental activity was conducted in two (2) phases: a pre-demo pilot phase and a sub-sequent final pilot phase. The final pilot was preceded by a pre-demo which aimed to verify the technologies and ensure safe and reliable operation in the human context of the final pilot, taking into account also different environmental conditions. The SHOW final pilot study aimed to serve as a small-scale evaluation or a real-life condition service, in order to then allow to conduct a larger study, which, however, goes beyond the objectives of SHOW.

The structure of the chapter is as follows. Section 2 presents the city of Trikala and the typical logistics services currently operating in the pedestrian area of the city centre. It also introduces the small AVs and the new logistics services that operated in the pedestrian area, defining also the research questions for the specific pilot operations. Section 3 describes the challenges faced and the proposed solutions. Section 4 presents the lessons learned from the experience and key suggestions for future similar applications. Section 5 concludes the chapter.

## 2 Content and the Research Questions

### 2.1 The Trikala Pilot Site and the Typical Logistics Service

The population of Trikala for 2023 is 81,849 inhabitants. The municipal area of Trikala is 607.59 km². In the city centre, there is a large pedestrian area, where many shops, including clothing shops and cafés, can be found. Currently, each brand has a warehouse outside the pedestrian area and uses its own conventional vans or heavy

**Fig. 1** Left: Manual loading/unloading of packages from heavy vehicles for merchants. Right: Packages left outside the merchant's shops before working hours. *Source* Authors' pictures

vehicles to make deliveries from its warehouses at the outskirts of the city to its shops, including those in the urban pedestrian area.

In Trikala, for the merchants in the mentioned pedestrian areas, and according to the local government rules, the heavy vehicles are allowed to deliver the goods to these merchants only in the early morning hours from 6 to 8 am. For this reason, the distributor leaves all the packages outside the merchant shops even before their scheduled business hours. The packages are left unprotected from theft and damage due to human or natural causes (e.g., rain). Figure 1 shows how the packages are currently loaded/unloaded and left outside the merchant shops.

## 2.2 The Small AVs: Yape Delivery Robots

Among the emerging automated urban freight delivery concepts (Cregger 2020), small AVs have been chosen to be used in the logistics network as the areas identified for the ser-vices are pedestrian areas. The selected small AV is Yape. It is characterized by com-pact size: it weighs 50 kg and can carry up to 10 kg of load and can go up to 9.6 km/h. Yape is equipped with: one lidar sensor to detect obstacles and detect the path to move autonomously, Li-ion battery which makes it to operate 8 h with-out interruptions, 3 HD and 1 stereo camera that allow the operator to monitor the delivery operations of Yape. Yape can go up to 12° of inclination.

Yape is part of a system which allows logistics services to be operated and is composed of:

- Fleet of Yapes that perform deliveries.

- Admin Platform that allows to manage the maps of the site where Yape navigates. The admin platform also provides the latest release of the Control Room and Yape software.
- Control Room that allows the operator to monitor the fleet during the delivery and, if needed, allows the operator to take control and manage critical operations.
- Delivery Platform, also accessed by an application installed on the user's smartphone, that allows the user to place a delivery request (on-demand de-livery service) and monitor the status of the delivery.

Yapes are commercial products, of SAE level 4 and Technology Readiness Level 7 (Cregger 2020). They are normally rented to those who need to make deliveries between a fixed origin and several destinations. The supply of the vehicles takes place together with the construction of the virtual map of the area where the delivery service will take place. In this way, those operating the delivery service, via the Delivery Platform, can select the delivery destination point on the virtual map from time to time. This is normally the only parameter that can be controlled.

For the SHOW project, some Yape's functions were changed and new functions were added. First, since it was necessary to define a multi-stop parcel delivery ser-vice, APIs were developed to bypass the Delivery Platform and chain together multiple single-stop deliveries. A GPS system was also integrated to improve the localization of Yapes on the virtual map. Moreover, new functions were added in relation to safety: the possibility to change the maximum speed of Yape through the Control Room and the possibility to save, on a memory card installed on Yape, high quality images captured by the Yape sensors. These images were used to verify the safety of pedestrians' and cyclists' interactions with the Yapes in the shared-use public environment.

According to the system requirements, the Control Room laptop was connected through Ethernet to the network. The droids were connected through their internal modem to 5G network. Moreover, each Yape was provided with a 4G SIM card.

The delivery service is defined in three stages, described in the following.

1. Delivery request: the user (receiver) requires a delivery through the delivery platform. If it is accepted, it will be recorded in the control room database. An identification number is provided for the related sender and receiver.
2. Package loading, depicted in Fig. 2: the Sender scans the QR code located on the droid's lid. Upon successful authentication the lid unlocks itself so that parcels can be placed in the load compartment. Once the sender confirms the loading through the app, the droid resumes the navigation towards the receiver.
3. Package Delivery, depicted in Fig. 2: the Receiver scans the QR code. Upon successful authentication of the user, the lid unlocks itself and the parcels can be collected. Once the receiver confirms the collection through the app, the droid navigates back to the sender.

A mission may consist of one or more trips. Each trip has an origin (sender) and a destination (receiver). A trip is successful if Yape reaches the receiver and unlocks the lid after successful authentication of the receiver. If, on the other hand, a trip is

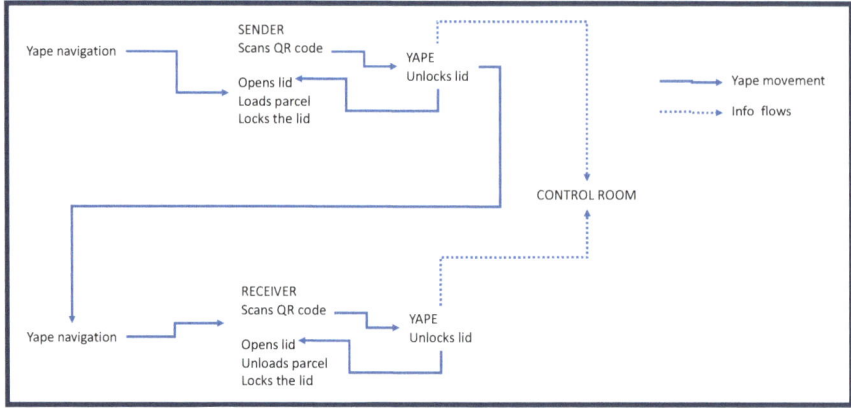

**Fig. 2** Package loading and delivery process. *Source* Authors' own representation

not successful, this means that a problem has occurred. In this case, the trip can be retried or the trip cancelled or the entire mission can be aborted (mission aborted).

The state of Yape can take on the following values: *Abort_Requested*, *Continue_Requested*, *Reached*, *Lid_Opened*, *Lid_Closed*, *Ongoing*, *Lid_Open_Requested*, *Completed*. *Abort_Requested* status is recorded when the control room operator aborts the delivery due to network issues occurred while performing the deliveries. *Continue_Requested* status is recorded when the operator tries to override the open/close lid process during the delivery; this occurred because of network interruption. *Reached* status is recorded when the Yape reaches the destination point and due to new user interaction difficulties, the delivery process cannot proceed to the next steps. *Lid_Opened* and *Lid_Closed* statuses are recorded when the Yape has reached the destination point and lid is opened or closed, and due to server or app issues it does not proceed to next steps of the delivery. *Lid_Open_Requested* status is recorded when the lid open command is requested by the operator or by the user and the delivery process does not proceed to next steps. This error usually happens when the new user does not perform lid operations properly. Alternatively, this status is usually recorded when the Yape is performing the delivery by moving to the destination point and it is interrupted by a non-user pressing the emergency button on the Yape. *Completed* status is recorded when the trip is successfully completed.

## *2.3   The Proposed Logistics Service Operated by Small AVs*

Three types of logistics services have been designed, as follows:

a. **Postcards collection** (occurring only in the pre-demo phase), operated as a launch event during the City Christmas festival, to introduce automated mobility to citizens and to make a first technological set-up for subsequent implementation of the logistic services in the pedestrian areas in the city centre. This first trial also allowed the operation of AVs in an outdoor environment under different weather conditions to be tested.

b. **Newspapers delivery** to 30 shops in the pedestrian area in the city centre (occurring in both the pre-demo and final pilot phases);
c. **Coffee remainders collection** from 2 cafés in the pedestrian area in the city centre (occurring in both the pre-demo and final pilot phases).

The original idea was to define a logistics centre in the urban area and to channel all deliveries destined for the pedestrian area from there. In this logistics centre, the following activities would be carried out: unloading of goods arriving there with conventional vehicles, storage, packing and consolidation of Yapes for distribution in the urban area. According to Settey et al. (2021) by inserting logistics centres into supply chains, the number of connections between suppliers and their respective receivers is significantly reduced (when comparing number of connections between each supplier and each receiver) as well as the number of transportation operations and the required size of the vehicle fleet.

However, two main problems prevented the implementation of a logistics centre as described above. The first problem is related to the fact that Yape has the following accessibility problems:

a. it can only move on smooth paths that are accessible to wheelchairs;
b. it has not been authorized to circulate on roads in mixed traffic with vehicular flows. It would have been able to move on cycling paths under mixed traffic conditions with bicycles, but the following problems would have remained:

- crossing road intersections and the associated need to make traffic light systems intelligent so that they could "talk" with the Yapes;
- the assurance of the safety of cyclists, which resulted particularly critical from the empirical research carried out in a protected environment (Mattas et al. 2021).

The second major problem is the load capacity of the Yapes. Since Yapes have important constraints on transportable weight and volume, not all parcels destined to the pedestrian area have the size/weight suitable to be transported by one of them.

The first problem results in strong constraints for the location of the logistics centre, which must therefore be in the pedestrian area, and the second problem results in the need for a heterogeneous fleet where, in addition to the Yapes, larger-capacity vehicles are used for larger packages, which cannot be transported by Yapes.

Therefore, a warehouse was chosen in the pedestrian area from which the distribution with the Yape fleet started. Typologies of products were chosen to be collected and stored in the warehouse, and then distributed with the Yapes, being suitable in weight and volume for the loading capacity of the Yapes.

Further problems have been raised by the stakeholders involved in the current supply chains: the goods delivered and stored in the new warehouse had to be purchased or insured for storage and transport. So as to avoid facing these new costs, it was decided to distribute promotional free goods with the fleet of Yapes.

A new warehouse has been built in the pedestrian area near the municipality office. The new warehouse was built to be accessible by the Yapes. All the five Yapes were stored, recharged and serviced in the warehouse. The operator defined the trips in the warehouse, and then, in case of delivery trips, s/he loaded packages into the Yapes which were missioned to the base point to start the deliveries; in case of col-lection trips, when the Yapes return to the warehouse, the operator unloads goods from the Yapes. Delivery/collection services are allocated to the first available Yape.

### 2.3.1 Logistics Service for Postcards Collection in Exhibition Area

The launch event of the Trikala experimentation within SHOW was organized during the Christmas festival that is organized every year in Trikala during the Christmas season. In the exhibition area, Yapes collected Christmas postcards (written by the children) from the post office and delivered them to Santa's house. The length of the round trip was approximately 500 m. The warehouse/control room was located in a railway carriage (part of the permanent structures of the exhibition space) near the Yape's route. In the warehouse, the operator scheduled Yape's journeys and sent the Yape to the post office; here an elf loaded the postcards into the Yape. Yape proceeded autonomously to Santa's house. When Yape arrived, Santa unloaded the postcards from the Yape. The Yape then returned to the warehouse.

### 2.3.2 Logistics Service for Newspapers Distribution to City Centre Shops

Daily, 30 copies of a newspaper were delivered to the warehouse by the journal editor. These newspapers were offered for free to the shops as a promotional introduction of automated mobility to Trikala city.

The mission performed by each Yape was a round multi-stop delivery journey: warehouse $\rightarrow$ shop1 $\rightarrow$ ... $->$ shop I $\rightarrow$... $->$ shop30 $\rightarrow$ warehouse. In this case, one mission was formed by 31 trips. The trips were created/requested by the operator (User) through their app. The 1st Sender was the operator working in the warehouse and the Receiver was the 1st shop-keeper. Then in the 2nd trip, the 1st shop-keeper be-came the Sender and the 2nd shop-keeper was the Receiver. In the last trip the Send-er was the 30th shop-keeper and the Receiver was the operator in the warehouse.

From the warehouse, each Yape travelled to shop1 and opened the lid as the shop-keeper took the newspaper and closed the lid. Then, Yape moved to successive shops and repeated the same procedure to deliver newspapers to all shops and travelled back to the warehouse. As both the delivery service and the newspapers were free, it did not matter if one of the shopkeepers collected more than one newspaper.

All the shops were along Asklepiou Street, as shown in Fig. 3. The multi-stop round delivery journey had a total length of about 300 m.

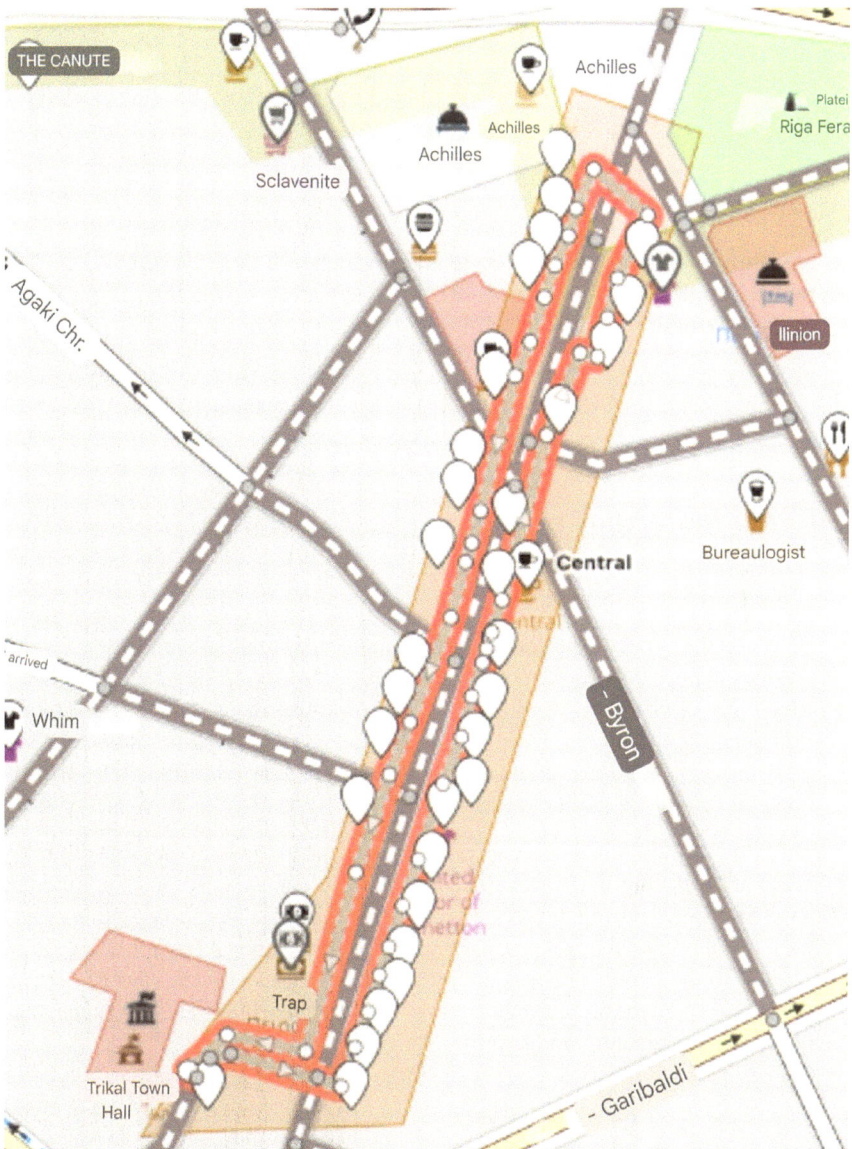

**Fig. 3** Newspapers delivery network layout. *Source* OpenStreetMap

### 2.3.3 Logistics Service for Coffee Remainders Collection in the City Centre

Two (2) cafés have been identified: they are located in Asklepiou and Vironos streets (Fig. 4). The objective was to collect the coffee residues from each café with Yapes and store them in the warehouse. From here, the coffee remainders were collected once a week by people working in BE-coop project (becoop-project.eu) in Greece. These coffee remainders are converted into pellets for heating purposes in children's schools and medical hospitals.

The service was on-demand. The Sender, that made the request (User), was the café-keeper and the Receiver was the operator in the warehouse. According to the café-keeper's requests, the mission performed by Yape resulted in.

- **one-stop journey**: warehouse → Café → warehouse (it is formed by 1 trip) OR
- **multi-stop journey**: warehouse → Café 1 → Café2 → warehouse (it is formed by 2 trips). In this case each café-keeper (Sender) made a request.

From the warehouse to Café 1 the distance is 160 m; from Café 1 to Café 2 the distance is 100 m; from Café 2 to the warehouse the distance is 270 m.

## 2.4   The Research Questions

The objective of the pilot operations in Trikala was to answer the following questions:

- what was the expected performances of the fleet of small AVs operating logistics services in pedestrian areas?
- what types of problems were expected and what was their frequency in Trikala's experience?
- what recommendations, based on the experience, can be formulated for future implementations of similar services?

The aim was therefore to test the capabilities of the individual Yape and of the fleet of Yapes in operating logistics services in specific contexts. To do this, numerous data were collected during the service operations. The data were analyzed in order to define a list of potential problems, to be understood and solved for future operations. According to the experience and the problems faced, general les-sons for delivery robots logistics operation in urban environment are presented. These lessons would facilitate the implementation of similar systems in the future.

## 3   Results and Challenges

During the logistics services operation, a database has been created. From this database the data for the analysis presented herein have been extracted. The database we refer to in the present analysis, has a number of rows equal to the number of per-formed trips. We selected two columns: the trip ID and the Yape's last status. From the status column, if there are no error and delivery is done, then it shows 'Completed'. If there was any kind of error, it shows the last status of the Yape.

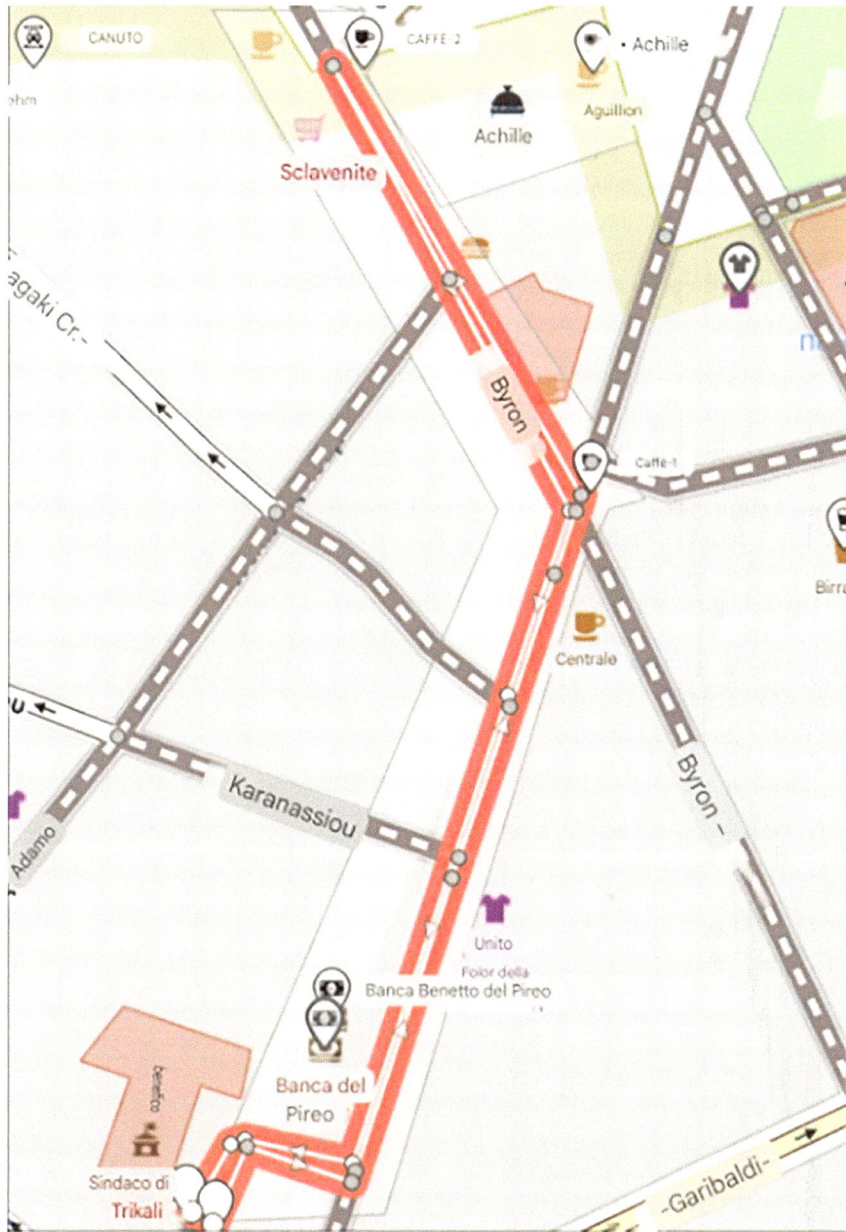

**Fig. 4** Coffee remainders collection network layout. *Source* OpenStreetMap

## 3.1  Operational Results

With regard to the distribution of newspapers, the pre-demo phase lasted 2 weeks: from 15 December 2022 to the end of December 2022 while the final pilot phase lasted approximately 1.5 months: from 3 January 2023 to 21 February 2023.

Regarding the collection of Café remainders, the pre-demo phase lasted approximately 1 week: from 24 January 2023 to 1 February 23 and the final pilot phase lasted ap-proximately 3 weeks: from 2 February 2023 to 21 February 2023. The operation hours for all the services were from 9 am to 6 pm. The following results refer to both pre-demo and final pilot phases. Figure 5 shows the number of successful trips daily performed by the 5 Yapes against time.

In the period 15th December 2022–21st February 2023, a total of 1,725 trips have been made by the 5 Yapes. Out of them, 1,496 trips were successful and 229 un-successful. Among the successful ones, there are: 203 postcard trips, 54 coffee trips, 1243 newspaper trips; among the unsuccessful ones, there are 57 postcard trips, 8 coffee trips, 160 newspaper trips. Figure 6 shows the cumulative number of successful trips against time. Figure 7 shows the distribution of the trips between the different types of services operated: 15.2% of the successful trips made were related to the collection of postcards in the exhibition area during the launch event. On the other hand, 84.4% of the trips were made in the pedestrian area in the city centre: in particular, 2.8% of the total trips were related to the collection of coffee residues and 82% to the delivery of newspapers.

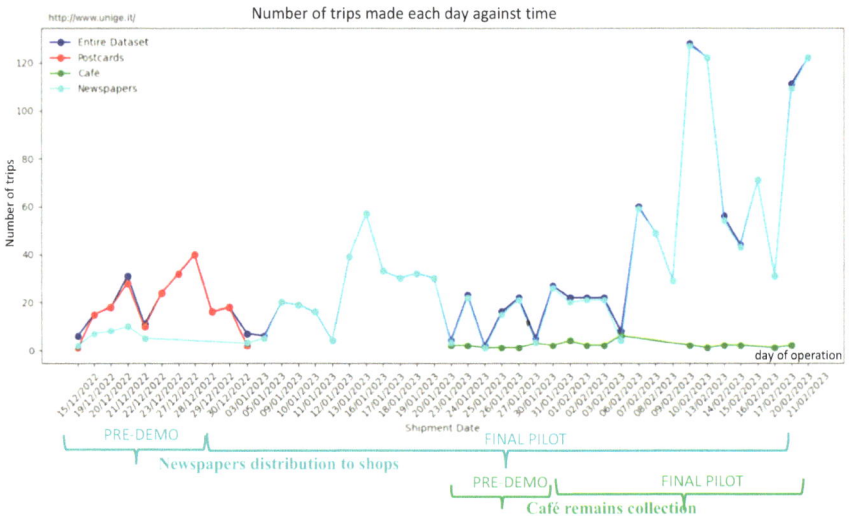

**Fig. 5**  Number of successful daily operated trips against time (3 months)

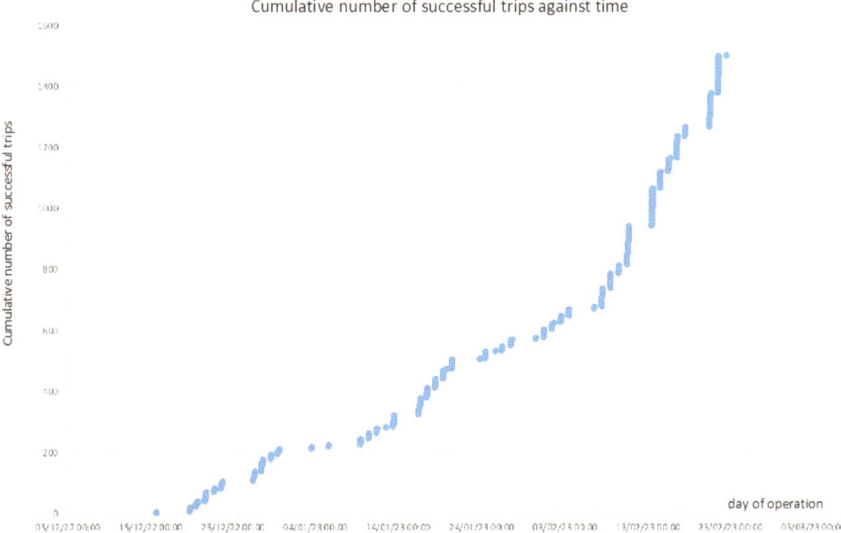

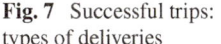

**Fig. 6** Cumulative number of successful trips performed by the 5 Yapes against time (3 months)

**Fig. 7** Successful trips: types of deliveries

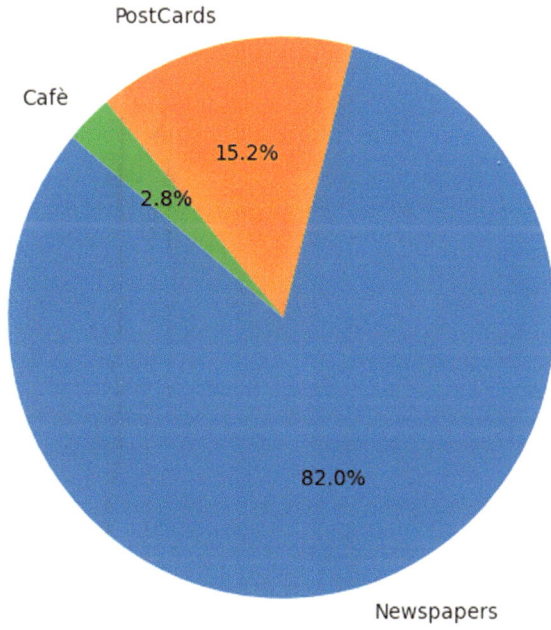

## 3.2  Operational Challenges and Mitigation Actions

If a trip is unsuccessful, it implies that a problem has occurred. The incurred problems have been classified into the following 4 typologies of incident/ error:

A.  **Data network issues**

This issue is identified with the recorded status: *Abort-Requested* or *Continue-Requested*. The control room cannot communicate properly to Yape: the operator cannot move Yape, cannot open/close the lid of the Yape but they still can abort the trip or abort the mission. However, Yape is properly operating.

The status of *Abort-Requested* was recorded 32 times; the status of *Continue-Requested* was recorded 4 times. The frequency of occurrence of each specific type of incident/ error is given by (1):

$$Frequency\ of\ type\ K\ challenge = \left( \frac{Total\ number\ of\ type\ K\ challenges\ that\ occurred}{Total\ number\ of\ challenges\ that\ occurred} \right) * 100 \tag{1}$$

For Data Network challenges, the above turns to: [(32+4)/229] * 100 = 15,7% of total number of challenges occurred.

The control room is connected by ethernet provided by the municipality office and worked properly. Problems occurred due to disconnections between Yapes and the Cloud. These disruptions occur when:

– there are too many users of the Trikala's public Wi-Fi;
– Yape passes from one Wi-Fi router to another one. Trikala's public Wi-Fi is created in such a way that a Wi-Fi router is installed on each light pole. Each light pole is located approximately 20 m from each other. While moving from one pole to another, Yape disconnects from the previous pole and connects to the up-coming pole. This activity creates a disruption of the network for almost 5 s.

To mitigate the above issue, each Yape was provided with 4G sim card. Alternatively, and if there were no time constraints, a new public Wi-Fi network with private access would also solve the issue.

B.  **Server and APP issues**

This issue is identified with the recorded status: *ArrivedToDestination=Reached*, *LidOpened, LidClosed*.

This problem typology takes place when Yape is at a stopping point and:

– Yape cannot communicate to the control room and therefore cannot send its up-dated status when its status changes due to an activity that has been correctly carried out. The server therefore does not give Yepe permission to proceed with the next activity, thus Yape does not proceed with the move. For instance, the lid has been closed or the Yape has reached its destination, but the lid is recorded as not have been closed, even if it is, or the Yape is recorded not to have reached its destination, even if it is; OR

– The sender's and/or receiver's mobile phones have a weak connection and the APP doesn't work. Therefore, the notification that QR code has been scanned cannot be sent to Yape. Therefore, the lid is not unlocked.

The *LidClosed* status was recorded 39 times, whilst the *LidOpened* status of was recorded 4 times; the *ArrivedToDestination* status was recorded 64 times. Thus, ac-cording to Formula 1, the frequency of occurrence of Server and App related challenges is equal to 46,7% of total number of challenges occurred.

The cause is related to a break in communication between Yape and the cloud or be-tween the cloud and the server in the control room or poor connection on the receiver/sender's mobile phones. The exchange of these communications is depicted in Fig. 8. Since each droid has been provided with 4G sim card, the only cause of server issues is related to the APP on receiver/sender's mobile phones.

This type of issues is mitigated by prompting the Yape users to use (whenever feasible) high speed internet or better Wi-Fi and to avoid attending calls while using the Yape's Application.

C. **Non-user interaction errors**

This type of issues is corresponding to the *Ongoing* recorded status. While moving, Yapes share space with pedestrians and cyclists. Out of curiosity many of them press the emergency button on the top of Yape (see Fig. 9). Pressing the emergency button causes Yape to shut down and the mission to be cancelled. However, in the control room Yape is still moving and the mission results on-going. It is necessary for the operator to reach Yape and physically re-activate it.

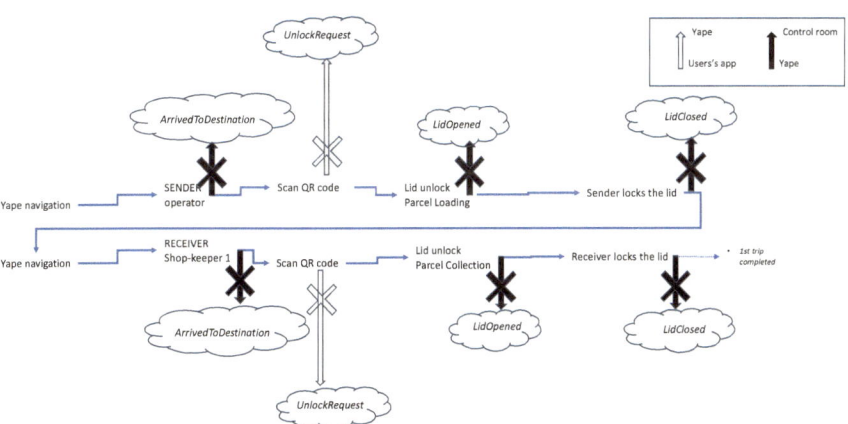

**Fig. 8** Communications and status notifications exchanged between Yape, the operator (via control room) and the user (via the app on his/her mobile phone). *Source* Authors' representation

**Fig. 9** Non-user interaction to the Yape while performing delivery operations. *Source* Authors' pictures

The *Ongoing* status was recorded 78 times. According to Formula 1, the frequency of occurrence of Non-User Interaction errors is equal to 34% of total number of challenges occurred.

To mitigate this, stickers were placed on Yapes mentioning in Greek: "Emergency button—Press Only in Emergency". After this, the occurrence of such incidents has been drastically reduced. The e-Trikala, local pilot partner, spread the news (in digital and physical social media and press) about Yape's operation and instructions on how to not obstruct the Yape's delivery service along with warnings to use the emergency button only in emergency situations.

D. **User interaction errors**

User interaction errors corresponded to the recorded *LidOpenRequested* status.

A typical case is as follows: Yape has reached a stop. The user has scanned the QR code, the app has notified Yape that QR code has been scanned, Yape has notified the remote controller that the lid is going to be unlocked, Yape unlocks the lid. However, Yape is not moving from there because the user doesn't perform lid opera-tion properly. The interaction between Yape and the user is shown in Fig. 10.

The *LidOpenRequested* status was recorded 8 times; the frequency of occurrence of *User interaction errors*, according to Formula 1, is equal to 3.5% of total number of challenges occurred.

To mitigate the above, training activities were organized by e-Trikala (in Greek language). Still, due to shift changes in many shops, this error still occurred. Thus, it was early decided that the training would be repeated once a week for the first 3 weeks of operation. Indeed, and after doing so, the issue was completely resolved.

**Fig. 10** User interaction mistakes with Yapes during logistical operations. *Source* Authors' pictures

# 4 Lessons Learned and Recommendations in View of Future Operations

Figure 11 refers to trips that failed and represents the frequency of occurrence of each type of incident/ error, both aggregated into the four aforementioned categories (on the left-hand side) and disaggregated into seven state value ranges (on the right).

- **1st lesson learned: the integration of radically new vehicles into existing logistics services is difficult and should not be underestimated**

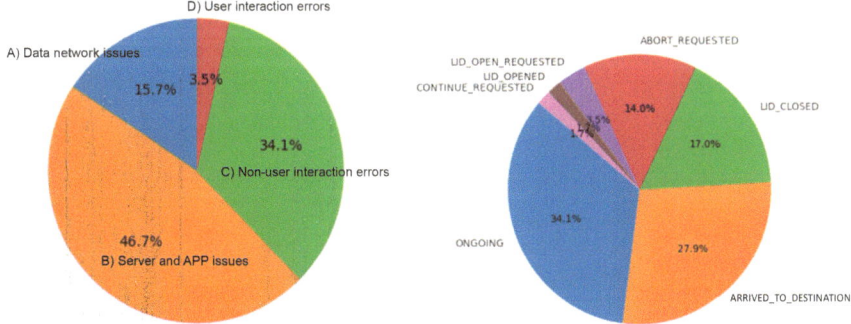

**Fig. 11** Left: Unsuccessful trips. Right: Frequency of occurrence of various types of incidents/ errors

The AVs that have been used (Yapes) have a different load capacity from the vehicles used in the typical logistics services of Trikala. The new vehicles could not replace the previous vehicles in their entirety, but they could have well replaced some of them to transport a portion of the goods: the smaller ones. They could therefore have been introduced to partially replace the existing logistics fleet.

In addition, Yapes have very stringent accessibility requirements due to the fact that they could not reach the existing consolidation centres outside the urban area. This made it necessary to introduce a new consolidation centre reachable by the new AVs. This would have led to an extra very costly load breakage: traditional vehicles would have had to transport the goods to the new centre and, from there, the Yapes would have taken care of the distribution in the pedestrian area.

*Recommendation 1: Anticipating Yapes and any new type of AVs for future logistics operations, and their accessibility or other requirements and restrictions, is crucial while identifying the serving consolidation centres of the logistics operation. Accessibility requirements concern both road intersections and the sections between two successive intersections.*

- **2nd lesson learned: the importance of having the trust and support of the population**

Experience has shown that more than 37% of the errors recorded are related to human—AV interaction, in particular about 34% of them to non-user interactions with Yapes and 3,5% of them to service user interactions with Yapes. As far as non-users are concerned, this is understandable since in the final pilot phase, the population circulated closely with the droids walking side by side in the pedestrian area. The population had to 'familiarise themselves with the droids' before letting them operate undisturbed in 'their' space. This result aligns with findings from Cepolina and Tyler (2004), Cepolina et al. 2017), in which it was studied how people move and react to different obstacles encountered in pedestrian environments. As for the users, in many occasions, they did not find easy to understand how to access the droids' storage compartment for loading/unloading operations.

*Recommendation 2: Training both the population and the users of the logistics service who will have to interact with the droids is crucial. The human–machine interface is again a critical aspect. The possibility for the remote controller to communicate directly, via the droid, with users and the population in case of doubts or information needs could facilitate human–machine interactions and make users more trusting of droids.*

- **3rd lesson learned: the importance of data network stability**

AVs need a stable data network in order to have a two-way communication with the control room. In our case, in order to have a stable network, it was necessary to provide each droid with a SIM card: this eliminated type A errors (Data network issues) completely.

*Recommendation 3: Ensuring a stable network immediately and radically reduces communication errors with the control room, which for our experience constituted 15% of the total errors. Good communication between the user's device and the droid is also necessary. This is more difficult to control and guarantee, depending on the operator and the network available to each user.*

## 5  Conclusions

The article is about a real implementation of logistics services operated with small AVs. One of the interesting aspects is that these services were operated in public urban areas, where circulation was not restricted to any kind of users, and therefore interactions between AVs and the population were very frequent. During the 3-month pilot, a lot of data was collected. From their processing, it was possible to assess the performance of the services operated and the nature of the problems that emerged, both of a technical and social nature. Clearly, in the pre-demo phases they occurred more frequently and towards the end of the experiment their frequency was markedly reduced as, especially in the pre-demo phases, solutions were adopted to solve the problems or at least mitigate their consequences. These mitigation actions are described in the paper in the hope that they will be useful in solving problems that may occur in future implementations of similar logistics services. Moreover, from the analysis of the collected data, it was possible to extrapolate suggestions that could avoid the future occurrence of the problems. These include the advice not to underestimate the difficulty of integrating radically new vehicles into existing logistics services; the importance of having the trust and support of the population and the importance of data network stability. Obviously, these suggestions are strongly linked to the specific type of AVs used as well as the specific operational context. However, it is SHOW Consortium belief that the experience gained can be of value to those who want to introduce small automated ground vehicles in logistics services over limited areas. The current experience gained is seen as an essential stepping stone towards progressively more efficient logistics operations that employ AVs.

**Acknowledgements**  The work presented in this manuscript has been funded by the SHOW project which has been made possible by funding from the European Union's Horizon 2020 Research and Innovation Programme Under Grant Agreement no. 875530.

## References

Akeb H, Moncef B, Durand B (2018) Building a collaborative solution in dense urban city settings to enhance parcel delivery: an effective crowd model in Paris. Transp Res Part E: Logist Transp Rev 119:223–233 (2018)
Altea ND (2022) Ricordate il robot Yape? Ora fa il postino in un quartiere di Milano [Online]. Available: https://www.wired.it/gallery/yape-droide-consegne-milano/. Accessed 18 April 2024
BBCNEWS (2018) Google's Wing delivery drones head to Europe [Online]. Available: https://www.bbc.co.uk/news/technology-46456694. Accessed 18 April 2024

Bouton S et al (2017) An integrated perspective on the future of mobility, Part 2: transforming urban delivery [Online]. Available: http://www.mckinsey.com/business-functions/sustainability-and-resource-productivity/our-insights/urban-commercial-transport-and-the-future-of-mobility?cid=eml-web. Accessed 18 April 2024

Cepolina EM (2016) The packages clustering optimisation in the logistics of the last mile freight distribution. Int J Simul Process Model 11(6):468–478

Cepolina EM, Tyler N (2004) Microscopic simulation of pedestrians in accessi-bility evaluation. Transp Plann Technol 27(3):145–180

Cepolina EM, Menichini F, Gonzalez Rojas P (2017) Pedestrian level of service: the impact of social groups on pedestrian flow characteristics. Int J Sustain Dev Plann 12(4):839–848

Cepolina EM, Cepolina F, Ferla G (2021) On line shopping and logistics: a fast dynamic vehicle routing algorithm for dealing with information evolution. In: International conference on harbour, maritime and multimodal logistics modelling and simulation, pp 27–36

Chen C, Demir E, Huang Y, Rongzu Q (2021) The adoption of self-driving delivery robots in last mile logistics. Transp Res Part E: Logist Transp Rev 146:102214

Cregger J, Machek E, Behan M, Epstein A, Lennertz T, Shaw J, Dopart K (2020) Emerging automated urban freight delivery concepts: state of the practice scan. Report N: FHWA-JPO-20-825, US Department of Transport

Deragni P (2019) Yape, il postino robot made in Italy, approda in Giappone [Online]. Available: https://www.wired.it/attualita/tech/2019/02/14/yape-giappone/. Accessed 18 April 2024

Gehrke SR, Phair CD, Russo BJ, Smaglik EJ (2023) Observed sidewalk autonomous delivery robot interactions with pedestrians and bicyclists. Transp Res Interdisc Perspect 18:4–8

Grolms M (2019) Autonomous shuttles and delivery robots [Online]. Available: https://www.advancedsciencenews.com/autonomous-shuttles-and-+delivery-robots/. Accessed 18 April 2024

Hawkins AJ (2024) Nuro gets a leg up from Arm in launching its third-generation delivery robot [Online]. Available: https://www.theverge.com/2024/2/22/24079434/nuro-robot-delivery-arm-power-efficiency-range. Accessed 18 April 2024

Heutger M, Kückelhaus M (2016) Logistic trends radar, DHL Customer Solution and Innovation, Germany [Online]. Available: https://www.dhl.com›full›dhl-trend-report-uav. Accessed 18 April 2024

Mattas K, Duboz L, Damy S, Cepolina EM, Silani E, Alonso-Raposo M, Ciuffo B (2021) Interaction between bicycles and automated delivery droids: results of the GNSS measurements and of the survey. In: TRB annual meeting

McFarland M (2019) UPS broke in to drone deliveries shuttling medical samples [Online]. Available: https://www.wtkr.com/2019/10/02/ups-broke-into-drone-deliveries-shuttling-medical-samples-in-north-carolina-now-its-ready-to-take-off. Accessed 18 April 2024

Molfino R, Zoppi M, Muscolo GG, Cepolina EM, Farina A, Nashashibi F, Pollard E, Dominguez JA (2015) An electro-mobility system for freight service in urban areas. Int J Electr Hybrid Veh 7(1):1–21

Reed S, Campbell A, Thomas B (2021) The value of autonomous vehicles for last-mile deliveries in urban environments. Manag Sci 68

ROBOTICS and AI (2019) Fraport trials AI-powered autonomous robot at Frankfurt Airport [Online]. Available: https://www.futuretravelexperience.com/2019/10/fraport-trials-ai-powered-autonomous-robot-at-frankfurt-airport/. Accessed 18 April 2024

Settey T, Gnap J, Beňová D, Pavličko M, Blažeková O (2021) The growth of E-commerce due to COVID-19 and the need for urban logistics centers using electric vehicles: bratislava case study. Sustainability 13:5357

Shields N (n.d.) China's Largest Courier is Starting Drone Deliveries [Online]. Available: http://www.businessinsider.com/chinas-largest-courier-to-start-drone-deliveries-2018-4. Accessed 18 April 2024

SHOW European Project (n.d.) SHOW (SHared automation Operating models for Worldwide adoption) project [Online]. Available: https://show-project.eu, Grant Agreement no. 875530. Accessed 18 April 2024

Vincent J, Gartenberg C (2019) Here's Amazon's new transforming Prime Air de-livery drone [Online]. Available: https://www.theverge.com/2019/6/5/18654044/amazon-prime-air-delivery-drone-new-design-safety-transforming-flight-video. Accessed 18 April 2024

# User Perspectives and Engagement

# Unlocking the Full Spectrum of User Perspectives on Automated Mobility Using the 'Supertesters' Method

**Dominik Schallauer, Aggelos Soteropoulos, Annika Dollinger, Alexander Mirnig, Peter Fröhlich, Allan Tengg, Alexander Moschig, Walter Prutej, and Petra Schoiswohl**

**Abstract** As vehicle automation advances, integrating automated vehicles into the existing transportation system is crucial, considering technical but also social factors. This chapter investigates two Austrian pilot sites, Graz and Pörtschach, by assessing user preferences through a novel "supertester" approach that included experiential elements as well as interviews, questionnaires and workshops. The supertester

D. Schallauer (✉) · A. Soteropoulos · A. Dollinger
Austria Tech Federal Agency for Technological Measures, Raimundgasse 1/6, Vienna, Austria
e-mail: dominik.schallauer@austriatech.at

A. Soteropoulos
e-mail: aggelos.soteropoulos@austriatech.at

A. Dollinger
e-mail: annika.dollinger@austriatech.at

A. Mirnig · P. Fröhlich
Center for Technology Experience, AIT Austrian Institute of Technology, Giefinggasse 4, Vienna, Austria
e-mail: Alexander.Mirnig@ait.ac.at

P. Fröhlich
e-mail: Peter.Froehlich@ait.ac.at

A. Tengg
Virtual Vehicle Research GmbH, Inffeldgasse 21a, Graz, Austria
e-mail: Allan.Tengg@v2c2.at

A. Moschig
AVL List GmbH, Hans-List-Platz 1, Graz, Austria
e-mail: Alexander.Moschig@avl.com

W. Prutej · P. Schoiswohl
Pdcp GmbH/SURAAA (Smart Urban Region Austria Alps Adriatic), Hauptstraße 204, Pörtschach, Austria
e-mail: walter.prutej@suraaa.at

P. Schoiswohl
e-mail: petra.schoiswohl@suraaa.at

approach is a within-subjects empirical method in which the same group of individuals experiences various use cases. Employing this approach allowed a comparative analysis across diverse settings, use cases, vehicle types and user perspectives. The study underscores the critical role of fundamental safety functions and the relation between different vehicle types and corresponding expectations of passengers.

**Keywords** Automated Vehicles (AVs) · User requirements · User experience · User expectations · Within-subjects user study · Vehicle type · Automated shuttle bus · Robo-taxi · Supertesters approach

## 1  Introduction

As the sophistication of vehicle automation technologies increases, so do the efforts to integrate automated vehicles (AVs) into the transport system and complement (and, in some cases, replace) existing mobility services. In this process, it is important to consider not only technical roadblocks, but also social aspects: How do citizens perceive and accept AVs, and what benefits can AV services bring to them? User perception depends on a multitude of factors and can vary in relation to population segment, vehicle type(s), purpose of travel, and many others (Millonig and Fröhlich 2018). Assessing such aspects comprehensively and comparatively in single-instance settings (i.e., through one single survey or one single ride with an AV) is, therefore, difficult.

This chapter investigates user perception and expectations for using automated public transport in different settings and with different automated mobility solutions in terms of use cases and vehicle types at two pilot sites in Austria, namely Graz in Styria and Pörtschach in Carinthia. The two sites cover a broad variety of relevant scenarios for automated mobility in Austria: urban, peri-urban, different vehicle and operating concepts. By introducing the "supertester" method, a within-subjects approach was employed wherein a consistent yet diverse group of individuals was recruited to experience both pilot sites. This approach ensured that the same set of users evaluated the different AV services across identical criteria. User feedback with regard to the AV services were gathered through questionnaires, interviews and workshops. Beyond the findings on user perception and expectation for AV services, the chapter discusses advantages and limitations from the application of the supertester methodology in comparison with other methods.

The chapter is structured as follows. Section 2 describes the methodology used to assess the user experience with regard to the different AV services. Section 3 presents the results regarding user experience and preferences with regard to the AV services, while Sect. 4 provides a discussion on the results of the study in relation to previous studies in the field and also presents limitations of the study. Finally, Sect. 5 entails a conclusion and identifies research gaps in the field and possible future avenues of research.

## 1.1 Related Work

Public acceptance, user experiences and preferences of AVs have been studied in several studies so far, with Bala et al. providing a comprehensive overview (Bala et al. 2023). Existing studies primarily investigated user experiences within tests of automated shuttle buses with a safety driver. In addition, recent studies also investigated the user experiences within the operation of other automated mobility solutions, i.e., robo-taxis, without a safety driver.

Schäfer and Altinsoy (2021), for example, analysed the user experience of 449 users within the test operation of two automated shuttle buses with safety drivers in Frankfurt, Germany by using an online-questionnaire. Results show that users assessed the on- and off-boarding and the environmental contribution very positive, while the driving speed (max. 18–20 km/h) was rated worse, i.e., only neutral to positive. Another aspect that was assessed negatively by the users was the lack of barrier-free access of the vehicle.

For Koppl in Austria, Zankl et al. (2017) investigated the user experience of 294 users within the trial operation of an automated shuttle bus with a safety driver also by using an online questionnaire. The majority of the participants stated that they found the trip very good (53%) or good (39%). Positive aspects named by the participants were the comfortable driving behaviour and the safe driving experience/ cautious driving style, while negative aspects encompassed the abrupt braking behaviour and jerky and influent driving style (i.e., frequent braking).

Stopher et al. (2021) surveyed 27 users of Waymo robo-taxis without safety drivers in Chandler, USA with regard to their user experience. Regarding the question whether they liked riding in Waymo AVs more than in a traditional ridesharing service or regular taxi, the majority of users strongly (41%) or somewhat (26%) agreed, while only 7% of the users strongly disagreed. More specifically, results show, that Waymo dominated the other services with respect to ride comfort and ease of getting into and out of vehicle, while Uber/Lyft was slightly better with respect to drop-off and pick-up.

Overall, there are several studies on user experience of AVs in the literature with mostly positive assessments by the users. Most studies employed online questionnaires after the rides and primarily focused on assessing user experiences and preferences for a specific AV service, i.e., automated shuttle buses or robo-taxis. While the results of the studies give hints on the user experience of specific AV services, a comparison between different automated mobility solutions is difficult, as users most likely did not experience other solutions. Comparing results for various solutions might be then biased because of different samples of users. So far, a study and respective specific method to assess the user experience from the same group of users encountering different AV services has not been conducted.

Therefore, this chapter proposes and investigates the supertesters approach, which facilitates a more comprehensive comparison of user experiences across various automated mobility solutions by applying within-subject assessments and including in-depth combined experiences, interviews, and workshops with the same users.

## 1.2 Relation to Other Activities of the SHOW Project

In the context of the European SHOW project (Horizon 2020, GA No 875530; https://www.show-project.eu/), a concise evaluation framework has been designed from the beginning (SHOW), tightly connected to the impact assessment framework of the project, defining the key research hypotheses for the shared AV services evaluation across its pilot sites. In this context, a series of data-driven and subjective-driven tools and processes have been employed. Along with performance data, subjective data are continuously sought and upon common digital survey tools, further translated to local pilot sites languages. Those have pursued to collective subjective feedback (upon closed and open queries) on different evaluation aspects (comfort, safety, usability and other essential factors) from all key stakeholders participating in the large-scale trials, namely the passengers of the AV services, the safety drivers of AVs, the operators, authorities, and other key stakeholders being specific to each local value chain and the other road users interacting with the AVs, including Vulnerable Road Users (VRUs). As inherent part of the evaluation and impact assessment framework, Multi-Actor Multi-Criteria Analysis (MAMCA) workshops (Assessment framework for final demonstration round 2022) have been conducted in each pilot site aiming to explore in a structured manner the view and assessment of stakeholders involved in the deployment of AV services across the pilot sites.

Still, as it has been revealed already in the first round of pilot evaluations in SHOW, the so-called "pre-demo phase of SHOW" aiming to serve as a rehearsal to the final pilot and open to selected internal project entities users (SHOW 2023), the employment of standardised subjective tools in massive numbers and in an inevitable not-controlled manner, did not allow the collection of an elaborate and context-specific qualitative insight of users (passengers) that would allow the deeper understanding of pros and cons as well as the impact of context-specific elements. In an equivalent way, it was also revealed from the first series of MAMCA workshops held (SHOW), that it is essential that the MAMCA workshops pursue, in parallel to the views of other stakeholders, specifically and most importantly more profound qualitative feedback of passengers—them being the end-users—who would be engaged to serve as supertesters in position to provide more thoughtful and concise evaluation of the AV services they have experienced.

As a result, and to fill-in the research-based gaps observed, the supertester approach was introduced in the project, with Austria being the first pilot site to apply this and establish it as an example process for other pilot sites following.

## 1.3 Objective

The aim of this chapter is to explore user experiences and preferences regarding different automated mobility solutions by introducing the novel supertester approach. By applying this within-subjects approach, we enable the same group of users, referred to as supertesters, to experience multiple use cases and types of AVs. This approach strives to provide a more comprehensive perspective on the mobility requirements for various use cases within automated public transport. In particular, we aim to address the following research questions:

- RQ1: What specific key improvement factors do participants consider as most crucial, when assessing the AV services?
- RQ2: How do certain characteristics of vehicle types and operational approaches across pilot sites impact users' perceptions and assessments?
- RQ3: What solutions and requirements emerge, when participants face additional challenges or problems during their journey?
- RQ4: Which advantages and limitations does the within-subject supertester approach provide across sites in comparison with other methods?

## 2 Methodology

### 2.1 The Pilot Sites

The pilot sites in Graz and Pörtschach are part of the "Austrian Triplet Mega Pilot" of the SHOW project. The sites use different types of AVs and operate in various traffic situations. Table 1 provides an overview of the features of the sites.

In compliance with the legal requirements for testing AVs in Austria Verordnung des Bundesministeriums für Klimaschutz, a safety operator is always present in the vehicles and the pilot sites provide their services free of charge. The vehicles used at the pilot sites are classified according to national regulations as an "automated vehicle for passenger transport" and an "automated minibus". The vehicle for passenger transport in Graz is allowed a maximum speed of 50 km/h, whereas the allowed maximum speed for the automated minibus in Pörtschach is 20 km/h. However, it is crucial to note that the actual speed permitted during testing is determined through a comprehensive analysis of the route and an assessment of potential risks. In numerous route segments, this determined speed is lower than the originally and generally specified maximum speed. During the test period, passengers must be seated in the designated seats in both sites. Individuals in a wheelchair and those with a stroller

**Table 1** Features of the pilot sites

| Pilot site | Vehicles | Vehicle type | SAE-level | Route length | Area |
|---|---|---|---|---|---|
| Graz | Ford Fusion and KIA e-Soul | Passenger Vehicle | L4 with safety operator | 2 km | City outskirts |
| Pörtschach (Carinthia) | Navya Arma DL4 | Shuttle | L4 with safety operator | 2.7 km | Village centre |

**Fig. 1** Graz pilot site: route and vehicles. *Source* Left—Map Data from OpenStreetMap; right—
authors' own picture

may be transported, provided appropriate safety measures are in place to ensure that
there is no increased danger for all vehicle occupants.

## Graz

At the pilot site, two SAE Level (L) 4 automated vehicles—a Ford Fusion Hybrid
and a Kia e-Soul—are operated on a two-kilometre-long route, connecting a mobility
hub to a shopping centre. For a major part of the route, the vehicles use a dedicated
bus lane. At the shopping centre, a short part of the route is driven manually by the
safety operator due to a complex intersection, parking cars and the presence of many
pedestrians. However, at the mobility hub, the vehicles are passing through the bus
terminal in automated mode—with many pedestrians crossing their way. No pre-
registration is needed for using the service; passengers can simply board the vehicles
and select their desired stop using a tablet computer. Each vehicle has a maximum
seating capacity of four (4) passengers (Fig. 1).

### Pörtschach (Carinthia)

The route in Pörtschach is 2.7 km long, providing a link between the train station,
town centre, hotels, shops and restaurants; with a total of 8 stops along the route.

   On the test day, the participants travelled from "Bahnhof" (train station) to
"Parkhotel" and "Wahlißwiese/Seeblick". These lakeside roads are relatively narrow
and shared with many pedestrians, bicyclists and parked cars. Given Pörtschach's
popularity as a tourist destination, the volume of pedestrians and cyclists along the
lakeside significantly increases on days with beautiful weather (Fig. 2).

   The Navya Arma DL4 shuttle has a capacity of eight (8) seated passengers. It is
also possible to secure wheelchairs and baby buggies in the shuttle, with access being
facilitated through an electrical ramp. The safety operator uses the built-in screen to
select a route and initiate the trip. In case a manual takeover is necessary, the operator
can assume control of the vehicle using a game console controller.

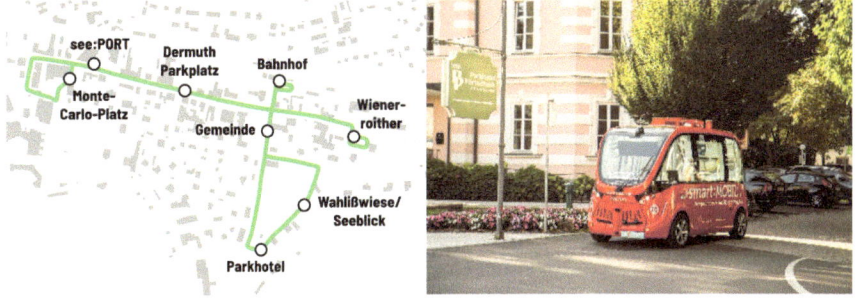

**Fig. 2** Pörtschach pilot site: route and vehicle. *Source* Left—Map Data from OpenStreetMap; right—authors' own picture

## 2.2 Participants

In total, 21 participants (10 females, 11 males) were recruited to experience the pilot sites on two dedicated test days (Fig. 3). The participation was promoted as becoming part of a supertester group, primarily through the social media platforms as well as the websites of AustriaTech and local project partners. To incentivise the participation, all travel expenses to reach and travel back and forth to the sites by public transport were covered. Additionally, catering on-site, including 30€ food vouchers per test day was included for all participants.

During their registration, 57% of the participants reported having interme-diate knowledge about automated mobility, while 43% claimed to have advanced knowledge. 64% of participants indicated, that they stay informed about recent developments in automated mobility.

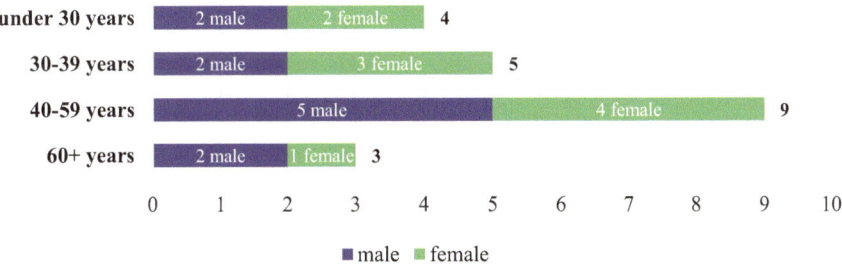

**Fig. 3** Participants by age group (n = 21)

## 2.3  Procedure on the Test days

**Test drives**

The test days took place end of June 2023 in Graz and end of July 2023 in Pörtschach. Each test day, i.e., the one in Graz and the one in Pörtschach started with a short introduction to the pilot site and the vehicles. During this session, all participants gave their informed consent for participation in the tests—all surveys were GDPR compliant and the results processing was anonymised. Subsequently, at each pilot site, all participants had one initial test drive. After the ride, participants were asked to rate their overall experience in a one-question-survey on a scale from 0 (not satisfied at all) to 100 (very satisfied) and filled in an additional acceptance survey (e.g., regarding reliability, usefulness or comfort). For this purpose, the SHOW evaluation protocol surveys mentioned in Sect. 1.2 have been used. Additionally, right after getting out of the vehicle, all participants were asked to share a few words about the experience (Fig. 4). The approach was to begin with a broad survey, aiming to capture general insights, and then dive into more comprehensive information and details during the workshop that followed the test drives.

For the next set of test drives, every participant was assigned a challenge card. These challenge cards were inspired by the CATAPULT project (2024) and were adapted for our tests. Each card presented a scenario, asking participants to engage in different hypothetical journey-related challenges (e.g. blockage of the vehicle by a spontaneous street demonstration). Some of the cards included the use of utensils like a wheelchair, crutches, glasses simulating visual impairments or noise-cancelling headphones. For the purpose of the challenges, participants were asked to pretend that there was no safety driver or any other personnel in the vehicle.

Following the second round of test drives, participants were once again asked for feedback regarding their experiences directly after getting out of the vehicle. Additionally, they were encouraged to exchange their insights during the following workshop with all other participants (Fig. 5).

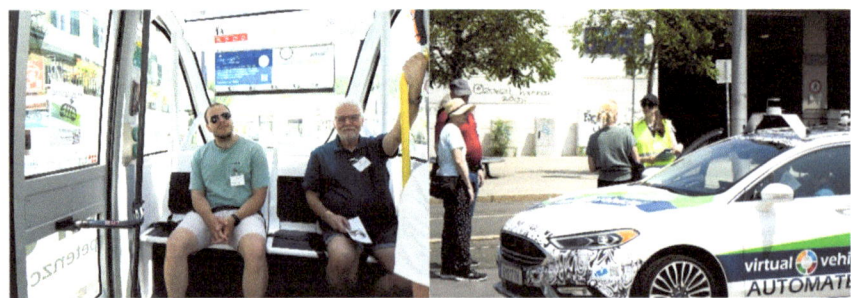

**Fig. 4** Passengers inside the shuttle in Pörtschach and participants giving feedback after the test drive in Graz. *Source* Authors' own pictures

**Fig. 5** Participants working in groups at the workshop. *Source* Authors' own pictures

## Workshop

Following the test drives, each test day concluded with a workshop following the same procedure:

- **Group Session**: In this segment of the workshop, participants formed small groups to discuss their experiences. Two flipchart posters were provided, each including a series of guiding questions.

Poster 1—Specific experience with challenge cards.

- What were the main challenges posed by the card or with the utensil?
- What needs to be changed or adjusted?
- Where were the challenges along the route?
- Were there challenges within the vehicle?

Challenge cards entailed specific events and situations that occur during the operation of the automated mobility solutions (e.g., technical problem, conflict between passengers) or specific characteristics a potential user has (e.g., hearing difficulty, broken leg, accompaniment of child).

Poster 2—General Experience.

- How satisfied were you with the ride in general?
- How did you experience the braking and acceleration behaviour of the vehicle?
- How did you perceive the speed of the vehicle?
- How did you perceive the accessibility?
- If this service was commercial, what would you need before, during, or after the trip? (With a focus on automation)
- Would you use this service again tomorrow? If not, what would need to change or be further improved?

Participants documented their insights by writing their responses on post-it notes and placing them in the corresponding sections of the flipchart poster.

- **Panel Session**: During this phase, all groups were asked to place their most note-worthy post-it notes onto a large pilot-site poster, positioned at the front of the room. One representative from each group presented the key takeaways.
- **Dot Voting**: To conclude the workshop, the moderators initiated a dot voting activity. Each participant received three dot points to mark the post-it notes they considered most important. Finally, the moderators summarized the topics receiving the highest number of dots.

## 3  Results

### 3.1  User Assessment of the AV Services

**Vehicle performance**

Following the rides, users were first asked to rate their overall satisfaction by moving a slider between 0 (not satisfied at all) and 100 (extremely satisfied). In both sites, in Graz and in Pörtschach the overall average satisfaction with the AV services was generally high, with a slightly higher rating observed for the robo-taxi on the initial test day in Graz (88.7) compared to the automated shuttle bus on the second test day in Pörtschach (80.4).

Figure 6 presents the evaluation of reliability, usefulness, speed, comfort and perceived safety of the AV services by supertesters using a scale of agreement to statements from 1 (complete disagreement) to 9 (complete agreement).

Reliability received a slightly higher assessment in Graz (6.9) than in Pörtschach (6.7). Additionally, participants rated usefulness more positively in general (above 7.5 in both Graz and Pörtschach, with minimal differences observed in the assessment

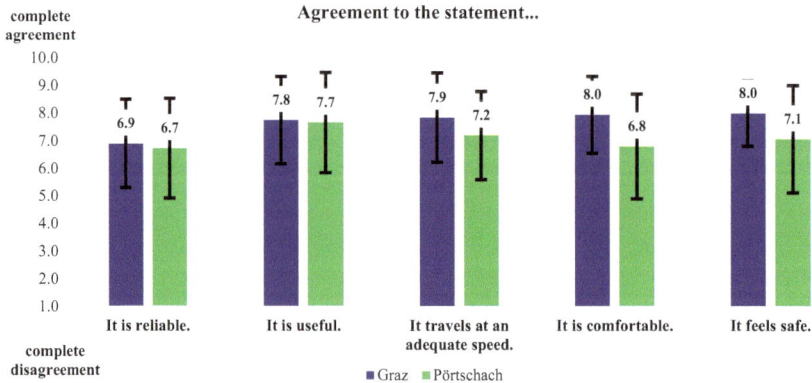

**Fig. 6** Assessment on reliability, usefulness, speed, comfort and perceived safety of the robo-taxi in Graz and the automated shuttle bus in Pörtschach by supertesters (Graz n = 20, Pörtschach n = 15)

of usefulness between the robo-taxi in Graz and automated shuttle bus in Pörtschach). Supertesters also rated speed, comfort and perceived safety higher for Graz compared to Pörtschach, with a more significant difference for comfort.

While the results on vehicle performance generally seem to paint a positive picture, statements and feedback gathered after the rides presents a more neutral perspective. In Graz, a participant highlighted the frequent interventions required from the safety driver, noting: *"Jerky, often braked strongly, very passive, safety driver had to intervene frequently."* In Pörtschach, one participant appeared positively surprised but, again, highlighted an issue with strong braking: *"Surprisingly smooth and perceived it similar to human driving behaviour. A strong braking was noted and perceived as negative, but overall still considered acceptable, as it was a singular occurrence."* In relation to perceived safety, one participant mentioned several challenges encountered during the journey in Pörtschach. Despite these difficulties, the participant acknowledged feeling safe overall, noting: *"Inconsistent; the shuttle had difficulty orienting itself but handled challenging driving situations well. It often came to a stop without apparent reason. Feels safe. Technology still needs improvement."*

During the workshop, users were additionally asked about their general experience with the AV services. In Graz, users assessed the rides overall as good, especially for the Ford Fusion vehicle, but it was mentioned that in some situations more feedback from the vehicle (e.g., information about arrival times or why the vehicle has stopped) would be needed to improve the experience. In addition, the ride especially with the Kia vehicle was assessed partly jerky and the way of driving rather slow and too defensive. In Pörtschach, overall, users were very satisfied with the ride as well as with the speed of the vehicle but mentioned that the braking was too rough.

**Willingness to pay**

Besides an assessment of the vehicle performance, participants also expressed their willingness to pay for the different AV services they experienced, i.e., the robo-taxi in Graz and the automated shuttle bus in Pörtschach. In detail, participants were asked how much they were willing to pay for a single 10 min ride of 2 km with the respective automated mobility solution experienced.

As illustrated in Fig. 7, participants expressed a significantly higher willingness to pay for a robo-taxi ride in Graz, averaging at 3.65 €, compared to the automated shuttle bus in Pörtschach, for which participants indicated a lower willingness to pay at only 1.39 €.

## 3.2   Prioritized Key Improvement Factors for Automated Mobility Solutions

Regarding factors that could improve the automated mobility solutions, i.e., improvement of existing features or desired options that are missing in the current AV services,

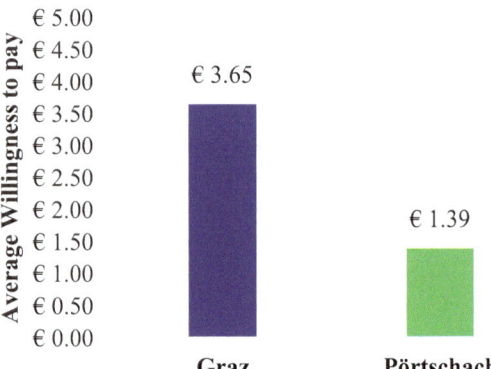

**Fig. 7** Assessment on the willingness-to-pay for the robo-taxi in Graz and the automated shuttle bus in Pörtschach by supertesters (Graz n = 20, Pörtschach n = 15)

for the robo-taxi in Graz, supertesters mentioned the need for existence of an emergency call centre and an emergency button as well as communication of instructions in abnormal situations (e.g. what are the next actions, when can I get off the vehicle) and a speech module that complements the information on the screen but also allows for back-and-forth communication.

With regard to the automated shuttle bus in Pörtschach, most stated key improvement factors were an easier wheelchair attachment and an application for on-demand requests, emergency calls and offers. A similar application was proposed in Graz, although it did not receive as many votes (3). For the shuttle bus, some participants noted, that the emergency door release was not visible to them, and that it was difficult for them to see the information on the central information screen from their seated position. They therefore proposed that the information could be displayed on the screen of their phone (Fig. 8).

In light of the outcomes, it is important to consider that some of these factors emerged in response to the challenge cards outlined in Sect. 2.3, under the assumption that there is no safety driver or other personnel in the vehicle.

## 3.3   Comparison of Various Vehicle Types and Operational Approaches

Table 2 summarises the insights gathered during the workshop that was conducted in addition to the survey after the test rides, outlining the participants' perceptions of the rides. Although overall satisfaction was consistently positive in both cases,

**Key improvement factors for AV services (top 4)**

**Number of votes by Supertesters**

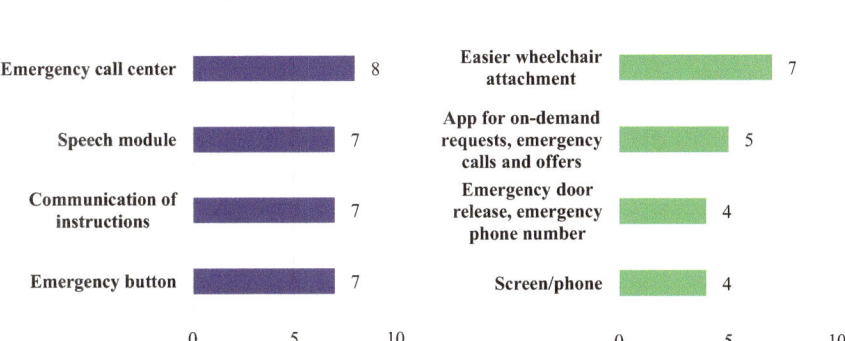

**Fig. 8** Key improvement factors mentioned by supertesters for the robo-taxi in Graz and the automated shuttle bus in Pörtschach (Graz n = 20, Pörtschach n = 15)

**Table 2** Perception of specific factors related to the rides

|  | Graz (robo-taxi) | Pörtschach (automated shuttle bus) |
|---|---|---|
| General satisfaction with the ride | Overall good, especially for the Ford Fusion vehicle | Very good, Fluent |
| Experience of braking and acceleration behaviour | Good for the Ford Fusion vehicle, partly jerky for the Kia vehicles | Acceleration fluent, but braking was too rough |
| Perception of the speed | Rather slow and too defensive | Good, Adjusted for the situation |
| Perception of the accessibility | Not accessible | Accessibility for wheelchair-users was assessed positively, practicable with ramp |

differences in braking, speed, and accessibility emerged across various vehicle types. For an overview of the actual vehicle performance data from the test days, refer to Table 3.

## 3.4 Issues Revealed Through Challenge Cards

During the second test drive, every participant received a challenge card, as described in Sect. 2.3. The cards included induced mobility restrictions and hypothetical journey related challenges (e.g. blockage of the vehicle by a spontaneous street demonstration). The experience of participants was then collected in the workshop and the results of the discussion are summarised in Table 4.

**Table 3** Actual vehicle performance data from supertester-day

| Vehicle performance data | |
| --- | --- |
| Graz (robo-taxi) route length: 2 km | Pörtschach (automated shuttle bus) route length: 2.7 km |
| Average speed: 20 km/h<br>Highest speed: 30 km/h<br>Hard braking events >3 m/s$^2$: 0<br>Operator takeover: Ford Fusion = 0, Kia e-Soul = 1–2 per ride | Average speed: 11 km/h<br>Highest speed: 14,4 km/h<br>Hard braking events >3 m/s$^2$: 2<br>Operator takeover: avg. 1 per ride |

**Table 4** Issues revealed through challenge cards

| Graz (robo-taxi) | Pörtschach (automated shuttle bus) |
| --- | --- |
| – Limited access to emergency stop button for passengers<br>– Absence of emergency number<br>– Lack of visual or audible information about the situation and planned actions<br>– Limited seat capacity and space<br>– Difficulty of getting in and out of the vehicle | – Limited visibility of emergency door release<br>– Absence of emergency number<br>– Limited width of seats<br>– Complex wheelchair securing mechanism<br>– Button to extend the ramp not easily reachable<br>– High incline of the ramp at stops with significant height differences to the curb or stops lacking a curb |

## 3.5  Analysis of Results

In the following, the findings from all test sites in relation to each of the initially posed research questions are analysed.

The key improvement factors for the AV services mentioned by most of the supertesters (RQ1) were related to safety and related features, where elementary safety functions (e.g., emergency stop buttons, emergency call functionality, manually operated emergency door or opening) were highlighted with increased emphasis in their necessity for vehicles without a safety operator present in the cabin, regardless of whether someone is alone or sharing the shuttle or robo-taxi with other passengers. An additional key improvement factor highlighted especially for the robo-taxi in Graz was the absence of visual information. Especially in abnormal situations, participants expected the vehicle to provide them with more information about the situation and planned actions.

Apart from these general implications, the workshop revealed some additional insights and requirements relative to their scenarios and vehicle types (RQ2). While a strong preference for audio-visual communication was expressed at both Graz and Pörtschach, it was particularly in Graz that the desired type of communication was described to be similar to an interaction with a taxi driver, implying more of a back-and-forth communication style.

Another interesting difference was the perceived speed: While the vehicles in Graz drove at almost double the speed of the one in Pörtschach, the Pörtschach-vehicle was perceived as driving at mostly the appropriate speed. The Graz-vehicles, however, were perceived as too slow overall—although these issues were only observable in the discussions of the users in the workshop, but not in the survey results. It is likely that here, too, the vehicle type shaped the participants' expectations towards what they would expect from a taxi or regular passenger vehicle, whereas the Pörtschach-vehicle was perceived more similar to a bus, where lower speed is expected and accepted as a result.

Regarding issues revealed by the challenge cards (RQ3), results indicate that ramps must be able to extend fully automatically and the button or other interaction device to do so must not be at a position where it cannot be easily reached while sitting or standing. Positioning a wheelchair in a low-capacity automated shuttle is tricky and requires a solution that enables easy fixation of the wheelchair (e.g., a hook or clamp on the inner wall and one additional strap for fixation), as well as a solution that allows regular seats to flexibly be modified to accommodate wheelchair users—a simple solution to the latter would be foldable seats for all fixed seat positions. These can already be commonly found in traditional public transport means but have yet to be applied by design for automated shuttles but also new vehicles concepts for robo-taxis.

Safety and emergency features were identified to be of primary importance. Currently, in both the robo-taxi and the automated shuttle bus, the button for an emergency stop is easily within reach for the safety operator of the vehicle, however its accessibility is not optimal for all passengers. Some participants noted that the button was not even visible from their seated positions. Anyhow, participants expect to be able to stop the vehicle in case of emergency on their own accord and to not have to depend on (solely) the vehicle's decision making to determine what is and is not an emergency. Furthermore, there needs to be clear communication before *and* during an emergency with availability of information both visually and auditory, as well as provided by the vehicle and upon request by the passengers (be it via apps or a dedicated information terminal in the vehicle). The participants also expressed that there needs to be a way to contact a human operator or supervisor in case of necessity and that removing the human element fully from the communication loop is not acceptable, especially in emergency or exceptional situations.

The methodology of supertesters across sites represents advantages as well as limitations (RQ4). On the one hand, the supertester approach allowed for a detailed comparison of assessments with regard to overall satisfaction, perceived safety, reliability and usefulness of the robo-taxi and the automated shuttle bus. However, having the same participants using the various automated mobility solutions at different dates, challenges occurred with regard to the availability of the same participants at different days and test sites. While the number of participants in Graz was 20, it decreased to 15 in Pörtschach. 5 participants were unable to attend due to illness.

## 4    Discussion

In the following, we further discuss the obtained results in light of contextual factors and related research.

### 4.1    Safety First

The overall assessment of the safety of the automated mobility solutions in Graz and Pörtschach was quite high and is in line with previous studies with a positive assessment regarding the safe driving experience like Zankl et al. (2017) in which automated mobility solutions were tested with a safety driver.

Also, the essential feature requirements across both sites were identified to primarily be related to safety and emergency features. More mundane features, such as emergency buttons, manual door openers, and similar already exist in manually operated public transport, yet were confirmed to have just as high, if not increased, importance in fully automated contexts (CATAPULT project 2024; Mirnig et al. 2020). An interesting aspect in this regard is that the potentially assumed need for increased security and surveillance in automated transport (e.g., to protect against or pre-empt harassment) that has been mentioned in some earlier studies, i.e., by users of a full-length automated bus in Sweden (Nordhoff et al. 2020), was not perceived as a major factor of importance at either site. This is in line with a similar finding by Winter et al. (2019) based on tests with automated shuttle buses in the Netherlands. From our findings and the related literature, we surmise that these factors are not necessarily less important, but likely that they are not as important as of yet. Despite the relatively high level of confidence in the technology, it appears that automated public transport is not yet sufficiently mature and, above all, not sufficiently ubiquitous for passengers to be concerned with aspects beyond basic safety features and the certainty that the journey can be completed in the first place, especially with regard to connecting services and the certainty of being able to get on or off at the right stop when there is no human reference point on board. Especially in regard to the latter, human support was found to still be expected and needed—if not on-board, then via remote communication, so that interventions in cases of errors, emergencies, or simple deficiencies of the system can be made for seamless operations. Having some kind of human support, e.g., via remote communication will remain important for automated mobility services.

### 4.2    Design Aspects of the Vehicle and Related Expectations

The investigation of different vehicle types in different environments yielded interesting insights regarding the passengers' expectations in relation to behaviour and interaction with the vehicles. The shuttle, due to its vehicle design and operational similarity to a regular bus, was expected to be slower than regular traffic, its lower speed, therefore, being acceptable to the participants. Contrastingly, the robo-taxis were perceived as too slow, a nuance that contradicted the survey results (see Fig. 6) but was clearly evident in the workshop discussions. In reality, the robo-taxis were driving faster than the shuttles.

While studies focusing on a single vehicle would often identify mismatches in expected vs. actual speed mostly in relation to overly passive driving (Nordhoff et al. 2020), we found that the perception of appropriate vehicle behaviour in terms of speed was also greatly influenced by passenger expectations relative to the vehicle type and type of service in the context of which this vehicle operates: the passenger vehicle, operating in the context of a robo-taxi service, being differentiated to a public transport service, led to different expectations by participants who expected a more direct, if not one-to-one interaction with the vehicle—be it via an app, such as in modern taxi services (e.g., Uber, Lyft) or an in-vehicle interface once they have boarded. For the shuttle, on the other hand, we found a more passive interaction approach to be expected and accepted, where passengers need to be informed about schedules and stops as in a regular bus but do not necessarily require further direct interaction beyond ticketing or interaction in cases of emergency.

In this context, it becomes evident that AV services should adhere to established patterns of conventional public transport, when it comes to selecting vehicle types. For regular line operations, buses and shuttles (depending on the required capacity) are likely better suited than passenger vehicles due to eliciting more appropriate expectations, both in terms of speed and expected communication infrastructure. Robo-taxis and smaller capacity on-demand solutions will likely be more suited as single-passenger vehicles. Deploying the same vehicle type, such as an automated shuttle bus, for both line and on-demand services, poses the potential risk of conflicting with user expectations regarding speed and interaction in both scenarios. Additionally, leveraging a vehicle's design features—and the expectations associated with them—can be a strategic approach for deploying vehicles that convey a perception of lower speeds in environments where such a characteristic is valued by the community, such as zones with high pedestrian volumes or near schools.

While the findings do not yet allow for more detailed planning of the appropriate vehicle type for any given purpose or user group, we can conclude that a one-vehicle-fits-all approach is unlikely to fulfil users' expectations across the variety of possible contexts and that vehicle type can be used as a tool to both influence and fulfil user expectations, with a variety of vehicles being key to satisfy expectations on the wider spectrum.

The willingness to pay of the supertesters for the services offered was found to be significantly different in Graz and in Pörtschach. The average willingness to pay for a 10 min ride with the robo-taxi in Graz was €3.65, which was substantially higher than the willingness to pay for a 10 min shuttle ride in Pörtschach (€1.39). This is also in line with the aforementioned expectations of users, which appear to be based on similarities between existing solutions and related prices. For instance, the robo-taxi is perceived as similar to existing taxis, while the shuttle is seen as comparable to regular public transport buses. The current higher fares for taxis compared to buses in public transport in Graz and Pörtschach may influence user expectations regarding prices. It is noteworthy, that the willingness-to-pay for the robo-taxi in Graz may still be low when compared to existing prices of current services. Exact fares for Waymo are not available, as they utilise surge pricing based on demand, with higher prices during periods of higher demand (e.g., weekends). However, Cruise employed a base fare of $5 with additional charges of 90 cents per mile and 40 cents per minute (Winter et al. 2019).

Furthermore, participants highlighted challenges related to accessibility and service-oriented information. Key issues included the absence of audible and visual cues from the vehicle, particularly in emergency situations. One of the most frequently mentioned factors for improvement was an easier attachment of wheelchairs in the shuttle bus. Also, a study by Schäfer and Altinsoy (2021) revealed, that despite the overall high acceptance among users, negative assessments of the automated shuttle bus were primarily associated with its lack of accessibility, specifically the absence of barrier-free access. However, overall, it seems that problems identified by the participants were mostly related to the more vehicle related design and its features and could be considered in future vehicle concepts.

Finally, it is crucial to emphasise that the aspects highlighted by the supertesters, such as the emergency call centre, speech module, easier wheelchair attachment, and emergency button, are particularly relevant in scenarios where the vehicle or service operates without a safety operator on board. It is also important to acknowledge that the vehicles tested do not represent the state-of-the-art for such services and that a safety operator was always present. Existing services such as Waymo or May Mobility robo-taxis, which operate without safety drivers, are more advanced and mostly encompass the features identified by the participants. For instance, Waymo collaborates with disability advocacy groups to ensure a safe and inclusive service (Dow 2023).

## 4.3   Limitations

While the results of this study deliver new insights by using a within-subjects approach, allowing the same group of users to experience multiple use cases and types of AVs, it also has some notable limitations. Firstly, the presence of a safety driver during all test drives may have influenced participant perceptions and behaviours, potentially leading to results that differ from real-world scenarios without a safety driver. Additionally, the recruitment attracted participants that were generally interested in automated mobility, possibly leading to a bias in their responses and experiences. While the intentional small sample size was chosen to facilitate in-depth discussions, it is important to note that the limited number of participants may restrict the generalisability of our findings.

# 5   Conclusion

In summary, the evaluation of the AV services in Graz and Pörtschach unveils specific challenges depending on vehicle type and use case, with a positive overall assessment aligning with previous studies. The key must-haves in automated mobility arise around typical safety and emergency features, with participants emphasising the importance of emergency buttons and manual door openers. The results also underscore the continuing relevance of human support, whether on-board or through remote communication.

The differentiation between shuttle and robo-taxi services reveal distinct passenger expectations. For instance, the robo-taxi experience is expected to be a more personalised, one-to-one interaction like a taxi service, whereas the shuttle service is expected to offer a more passive interaction approach, like traditional bus services. Therefore, it is recommended to align AV services and vehicle types with the established patterns of conventional public transport to meet users' expectations. Accessibility and service-oriented information was identified as a crucial aspect, with participants desiring more comprehensive information through both audible and visual cues, along with improving the wheelchair attachment in the shuttle bus.

While limitations exist in the study, including the influence of safety drivers and a potentially biased participant group, the findings underline the need for a nuanced approach to AV services. A one-vehicle-fits-all strategy is unlikely to meet user expectations across diverse contexts.

Moving forward, the supertester method could be further refined. For instance, future studies could benefit from exploring real-world scenarios without a safety driver, providing a more authentic understanding of user perceptions and behaviours. Additionally, efforts to diversify participant recruitment beyond those initially interested in automated mobility would contribute to a more comprehensive and unbiased evaluation. Expanding the deliberately small sample size for in-depth discussions with a broader participant pool could enhance the generalisability of findings.

In conclusion, the study has shown that meeting passengers' expectations in automated mobility requires a careful integration of essential safety and communication features, ensuring accessibility and being aware of different expectations related to various service and vehicle types. The ongoing evolution and rollout of AV services demands for a comprehensive understanding of diverse passenger preferences and expectations.

**Acknowledgements** The results are part of the SHOW project, which has received funding from the European Union's Horizon 2020 research and innovation programme under grant agreement No 875530.

# References

Bala H, Anowar S, Chng S, Cheah L (2023) Review of studies on public acceptability and acceptance of shared autonomous mobility services: past, present and future. Transp Rev 1–27

CATAPULT Project (2024) Serious game. Project funded by Urban Europe. Accessed 22 April 2024 from https://catapultproject.eu/serious-game/

Dow J (2023) Waymo starts taking fares for SF robotaxis after state approval. Electrek. https://electrek.co/2023/08/16/waymo-starts-taking-fares-for-sf-robotaxis-after-state-approval/

Johansson M, Ekman F, Karlsson M, Strömberg H, Jonsson J, Faleke M (2023) Automation as an enabler: passengers' experience of travelling with a full-length automated bus and their expectations of a future public transport system. Transp Res Procedia 72:957–964

Macharis C, De Witte A, Turcksin L (2010) The Multi-Actor Multi-Criteria Analysis (MAMCA) application in the Flemish long-term decision making process on mobility and logistics. Transp Policy 17(5):303–311

Millonig A, Fröhlich P (2018) Where autonomous buses might and might not bridge the gaps in the 4 A's of public transport passenger needs: a review. In: Proceedings of the 10th international conference on automotive user interfaces and interactive vehicular applications, pp 291–297

Mirnig AG, Gärtner M, Füssl E (2020) Suppose your bus broke down and nobody came. Pers Ubiquit Comput 24:797–812

Nordhoff S, Stapel J, van Arem B, Happee R (2020) Passenger opinions of the perceived safety and interaction with automated shuttles: a test ride study with 'hidden'safety steward. Transp Res Part a: Policy Practice 138:508–524

Schäfer P, Altinsoy P (2021) Autonom am Mainkai. Nutzerakzeptanz und betriebliche Herausforderungen autonomer Shuttles in Frankfurt am Main. Frankfurt University of Applied Sciences

SHOW D9.3: Pilot experimental plans, KPIs definition and impact assessment framework for final demonstration round (2022). https://show-project.eu/wp-content/uploads/2022/10/D9.3-Pilot-experimental-plans-KPIs-definition-impact-assessment-framework-for-final-demonstration-round.pdf. Deliverable of the Horizon 2020 SHOW project, Grant Agreement No. 875530

SHOW D11.3: Pre-demo evaluation activities (2023). https://show-project.eu/wp-content/uploads/2023/03/SHOW-WP11-D-UIP-003-01_-_Pre-demo_evaluation_activities.pdf. Deliverable of the Horizon 2020 SHOW project, Grant Agreement No. 875530

Stopher P, Magassy TB, Pendyala RM, McAslan D, Arevalo FN, Miller T (2021) An evaluation of the valley metro–waymo automated vehicle ridechoice mobility on demand demonstration. department of transportation. federal transit administration. https://www.transit.dot.gov/sites/fta.dot.gov/files/2021-08/FTA-Report-No-0198.pdf

Verordnung des Bundesministers für Klimaschutz, Umwelt, Energie, Mobilität, Innovation und Technologie über Rahmenbedingungen für automatisiertes Fahren (Automatisiertes Fahren Verordnung–AutomatFahrV). https://www.ris.bka.gv.at/GeltendeFassung.wxe?Abfrage=Bundesnormen&Gesetzesnummer=20009740

Waymo (2024) Waymo accessibility network. Accessed 22 April 2024 from https://waymo.com/waymo-accessibility-network/

Winter K, Wien J, Molin E, Cats O, Morsink P, van Arem B (2019) Taking the self-driving bus: a passenger choice experiment. In: 2019 6th international conference on models and technologies for intelligent transportation systems (MT-ITS). Cracow, Poland, pp 1–8. https://doi.org/10.1109/MTITS.2019.8883310

Zankl C, Rehrl K, Digibus (2017) Erfahrungen mit dem ersten selbstfahrenden Shuttlebus auf öffentlichen Straßen in Österreich.Salzburg Research. https://www.salzburgresearch.at/wp-content/uploads/2018/04/Digibus_2017_Endbericht_final.pdf

# Societal Impacts of Automated Mobility for Public Transport: Insights from a Modified Delphi Study and Expert Interviews

**Víctor Ferran, Ignacio Magallón, and Paola Rodríguez**

**Abstract** This chapter reports findings from a study using a consensus method with an expert and stakeholder panel (n = 78) to analyse the societal implications of Cooperative, Connected and Automated Vehicles (CCAV) in public transport. To address the uncertainty about the wider societal impacts of CCAV, this study combines the results of a modified Delphi study, insights from interviews conducted at pilot sites deploying CCAV in real-word environment, and expert interviews from various CCAM-focused European initiatives. The modified Delphi study assesses direct consequences of CCAV such as accessibility and equity of public transport, user-perceived safety, and the impact on job creation/destruction and re-skilling as well as indirect effects of CCAV such as variation in house prices. These impacts are assessed within four scenarios related to different services and business models being deployed and tested across Europe in the Horizon 2020 SHOW project (GA No 875530).

**Keywords** Shared automated mobility · Societal impact · Modified Delphi · Re-skilling · Accessibility · Public transport · Equity · Jobs impact

## 1 Introduction

Cooperative, Connected and Automated Vehicles (CCAV) is developing rapidly, but there are still uncertainties about their impact on society. Over the past decade, major improvements in industry and research have increased the testing of automated vehicles on roads, even more so since 2023, when three U.S. cities allowed automated

V. Ferran (✉) · I. Magallón · P. Rodríguez
Bax & Company, Barcelona, Spain
e-mail: v.ferran@baxcompany.com

I. Magallón
e-mail: i.magallon@baxcompany.com

P. Rodríguez
e-mail: p.rodriguez@baxcompany.com

© The Author(s) 2025
H. Cornet and M. Gkemou (eds.), *Shared Mobility Revolution*, Lecture Notes in Mobility, https://doi.org/10.1007/978-3-031-71793-2_9

vehicles to operate on their streets, and numerous pilots have been running across Europe. However, the current stage of these pilots and operations is still not sufficient to assess the societal impact of CCAV.

This study, aimed at understanding the societal impacts of CCAV, is developed as part of the impact assessment within the SHOW[1] project. SHOW aims to support the deployment of CCAV in public transport by testing in real-life urban environment in 22 cities across Europe. This chapter focuses on the societal impact of CCAV by integrating the findings from the literature and from various European initiatives, results from a modified Delphi study and from expert interviews. The societal impacts addressed are accessibility to public transport, equity, housing prices, perceived safety, and the impact on jobs.

The Delphi method, originally developed by Dalkey and Helmer of the RAND Corporation in the 1950s to forecast the effects of technology on warfare (Dalkey and Helmer 1963), has since been widely used across various fields such as health, education, management, and environmental science. This chapter presents the findings of a study employing a modified Delphi method. Unlike the traditional approach of beginning with an open-ended questionnaire distributed to a panel of experts to solicit specific information, the modified Delphi technique begins with a predetermined set of items. These items may be derived from a range of sources, including relevant competency profiles, comprehensive literature reviews, and interviews with select content experts (Custer et al. 1999). This method allows for engagement with experts and stakeholders involved in public transport in a structured consensus-building process to identify and assess the multiple impacts of CCAV deployment on different aspects of society, such as accessibility and equity of public transport, user-perceived safety, land use impacts on house prices and impacts on jobs, including creation, destruction, and re-skilling.

The pilot site interviews within the SHOW project provide us with examples of the implementation of CCAV in a real-world environment. The direct experience and contact with passengers and value chain employees give us valuable insights into how these services affect the most immediate users. It is also an opportunity to validate the consensus reached in the Delphi. For this study, the sites of Monheim (pilot in a city centre in Germany) and Les Mureaux (pilot in the Ariane private site in France) were selected for interviews, because they have tested different types of services and have been running long enough to collect a sufficient amount of results (18 and 11 months respectively).

Additionally, the impact on jobs is explored building on recently completed European projects such as Skillful (2020),[2] WeTransform (2020),[3] and other studies carried out by the European Commission (2021).

---

[1] SHOW is a project funded by the European Union's Horizon 2020 under Grant Agreement number 875530, https://www.show-project.eu/.

[2] Skillful is a project funded by the European Union's Horizon 2020 Research and Innovation Programme under Grant Agreement number 723989, https://www.skillfulproject.eu/.

[3] WeTransform is a project funded by the European Union's Horizon 2020 Research and Innovation Programme under Grant Agreement number 101006900) https://www.wetransform-project.eu/.

## 2  State of the Art

The societal impacts of CCAV have been researched and discussed in the literature. Some of these are direct impacts of CCAV, while others are more indirect outcomes resulting from the broader societal and urban changes induced by CCAV technologies.

### 2.1  Accessibility to Public Transport

Accessibility is defined by Cohen and Cavoli as "the relative ease with which individuals are able to gain access to the locations, goods and experiences that are important to them" (Cohen and Cavoli 2019). This relates to the 15-min city concept, originally proposed by Carlos Moreno in 2016 (Moreno 2021), the principle that essential basic services should be accessible within 15 min by active mobility modes, which has been adapted to other time intervals and modes of transport, including public transport at region level, for example Ile-de-France Master plan, which aims to create coherent living basins within the "20-min region" concept (Schéma directeur de la Région Île-de-France (SDRIF-E) 2040).

Shared automated vehicles could improve accessibility, especially in peripheral areas, triggering further urban expansion. City centres could also be affected by a shift of parking to the periphery, and thus an increase in the density of economic activity in city centres (Milakis et al. 2017). In the long term, if the overall cost of moving is significantly reduced, this can lead to urban sprawl, with people moving further away from centres (Brown 2014).

### 2.2  Equity

Equity in transport has been approached from different angles and it still has different interpretations. For instance, Bruzzone et al. (2022) relates it to concepts such as justice, convergence, and fairness. In this chapter, we have addressed it including physical accessibility to the vehicles (i.e., boarding and alighting) (Whitmore 2022; Litman 2024), social integration, and community cohesion (Bruzzone et al. 2022), where public transport is argued to be a catalyst for this development (Litman 2024). Vehicle automation can have both positive and negative impacts on social equity, as it can bring social justice by enabling people with travel limitations, such as the elderly or persons with disabilities, to travel more easily (Harper et al. 2016), while it can create a negative effect, such as an induce of traffic demand of up to 14% of current non-drivers (Eby et al. 2016). To make sure that communities that need more robust transportation options are benefited by shared automated vehicles, policymakers must regularly solicit public input (Center 2023).

## 2.3  Housing Prices

Gelauff et al. (2017) show that the automation of public transport leads to a higher concentration of population in urban centres, while the automation of private cars leads to an increase in population in rural areas and suburbanisation. These changes in population distribution have a direct impact on housing prices, especially if they increase the density of the already most attractive places for people to live.

Still, housing prices can be affected by the type of services that automation can bring. The growth of automated vehicle-sharing services could reduce the need to build off-street parking, potentially increasing housing affordability (Milakis et al. 2017).

## 2.4  User Perceived Safety

A strong correlation is found between the perceived safety of automated vehicles (AVs) and the intention to use them (Koul and Eydgahi 2020). It is also revealed that road accidents involving AVs will affect the public perception and trust in using them (Zhang et al. 2024).

There are several studies suggesting that AVs could dramatically reduce the number of car injuries. For instance, recent studies from Waymo (Scanlon et al. 2022) in the U.S. claim a big increase of effectiveness and avoidance of 75% of collisions, leading to 93% reduction of serious injury risk. Still, one of the main barriers in achieving those safety goals, seems to be rather phycological and not technological (Shariff et al. 2017) suggesting that it is crucial to understand the recognition and the criticality of the factors affecting people's acceptance and, thus, AV adoption (Xu et al. 2018).

## 2.5  Impact on Jobs

Research suggests that shared vehicle automation will lead to changes in the type jobs and skills required in the transport sector, including the disappearance of some jobs and the emergence of others. The number of typical driver jobs will be reduced, while an increase in the number of computer specialists is expected, as well as new high-skilled jobs including ICT competences, e.g. in manufacturing, maintenance and transport-related tasks. The skills required for driving a vehicle will also change as automation gains full control of the vehicle, e.g. requiring more supervision and selective skills (Alonso Raposo et al. 2018; Ciuffo and Raposo 2019).

Measures to mitigate the negative impact of the deployment of AV services include the reskilling and upskilling of current workers. The International Transport Forum

(ITF) defines upskilling and reskilling as people's willingness to adopt new skills for their current (upskilling) or for a different (reskilling) job (ITF 2023).

However, it is worth noting that many European countries are experiencing a shortage of drivers in public transport that has escalated in recent years due to demographic changes, alterations in working conditions and the introduction of new technologies, and AVs could be a way to counterbalance the situation (ITF 2023; Okamoto 2019).

The reviewed literature reveals that research has been conducted on the topic, employing various methodologies. However, thus far, the impact of CCAV on society remains unclear, with studies often focusing on isolated aspects. Our study seeks to address this gap by integrating diverse methodologies and various perspectives in order to examine the broader societal implications of CCAV. To do so, this chapter aims to provide a consensual answer by experts and stakeholders for different scenarios deploying automated services in public transport through a modified Delphi study, which can then be discussed and interpreted by representatives from real-life urban pilots.

The modified Delphi method is one of the most relevant methodologies to assess problems of ambiguity and uncertainty. This approach consists of a survey conducted in an anonymised way in several rounds to a group of experts that are invited to participate through emails or online questionnaires. The experts give their independent opinions about each item. After each round, a report with the results from the previous round is sent to the experts for the next round, so that they can modify their opinions in order to increase the collective agreement, taking into account the results of the previous round. This process is repeated until a consensus is reached, or after a previously set number of rounds, that usually depends on the size of the sample and the consensus. With a small sample, no more than one round may be needed, however, a minimum of two rounds are usually required to allow valuable feedback and revision. The number of rounds may be set and "sacrificed" taking into account the continuity and guarantee of the study (Landeta 2006).

Giannarou and Zervas (2014) review various Delphi studies and compare methodologies and even if this method measures the consensus, there is no common practice regarding the statistical analysis of the results, with this approach varying from study to study (Landeta 2006). Hence, there are studies that measure agreement through frequency distributions and others using standard deviation or the interquartile range. For the analysis presented in the current manuscript, standard deviation is used as an accepted measure, and in order to reach a consensus, a standard deviation of less than 1.5 should be reached (Christie and Barela 2005).

## 3   Methodology

The societal impact of CCAV that has been researched in other projects, served as basis for this chapter and for the combination of the different methodologies used (modified Delphi, pilot site representative interviews and expert interviews). This study uses a combination of the above methods to assess the societal impacts of CCAV mentioned and provide local lessons learned, embedded in real-life demonstration, that could be generalised to other European cities. Figure 1 shows the process of the Delphi process followed for this study.

For this study two rounds have been required to reach consensus, with 78 participants in the first round and 40 participants in the second round. The first part of the survey served to collect data from the participants, including experts and stakeholders as shown in Fig. 2, including Public Transport Operators (PTO), city/region representatives, Original Equipment Manufacturers (OEMs), and research/academia along with their years of experience (Fig. 3). It is also denoted whether participants are SHOW partners or not (Fig. 4), whilst participants were asked to choose between the four different pre-selected scenarios (Fig. 5), that are described in Table 1.

Once the profile had been defined, the experts were asked to select one of the four (4) future scenarios tested across Europe, with the hypotheses of the services described in Table 1.

Participants have been asked to rate the different impacts of each service on a scale from −100 to 100% to determine whether the implementation of a CCAV service decreases/worsens (depending on the questions) or increases/improves compared to the current state, which is 0% and represents no impact. The questions having been included in the study are (Table 2).

The standard deviation of the answers has been used to measure consensus among the participants, and to calculate standard deviation the values −100 to 100 have been divided by 10 to match the literature examples scales. According to Christie

**Fig. 1** Delphi process in SHOW. *Source* Authors' representation based on Beiderbeck et al. (2021)

**Fig. 2** Sector

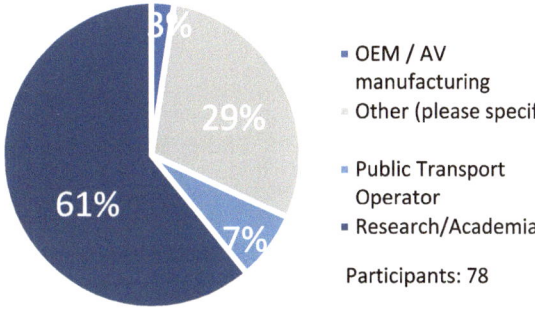

**Fig. 3** Years of experience in AV

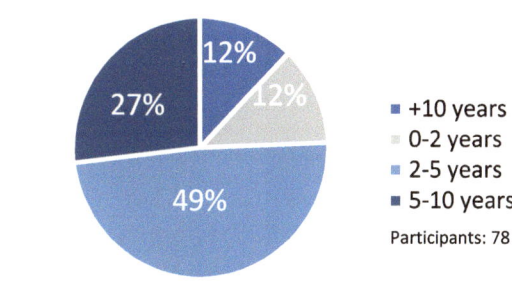

**Fig. 4** SHOW partner

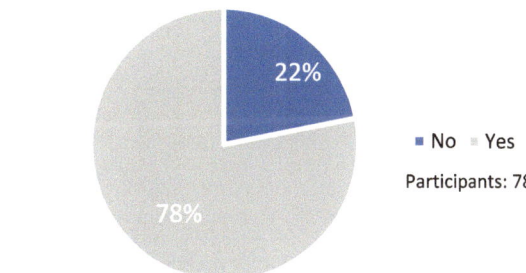

**Fig. 5** Scenario

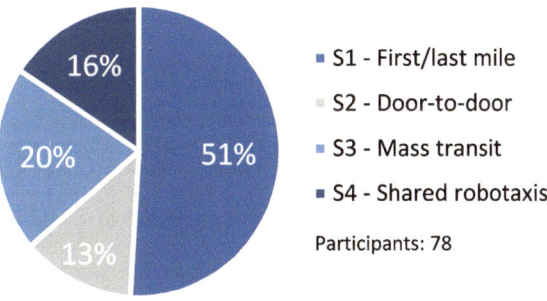

**Table 1**  Scenarios tested in SHOW

**Scenario 1** Automated shuttle(s) for first/last mile

This shared service acts as a feeder service to public transport for the first/last mile. In SHOW, these connections include hospital campuses, universities, school, and residential areas. The medium–low speed shuttles follow a fixed route to or from public transport stations with the possibility to implement on-demand stops or fixed stops. The service operates in parallel to high-capacity public transport. Depending on the area, the service operates on a fixed line with fixed stops or can serve as an on-demand service, where the user requests a pick-up at the nearest stop. When in full operation and given there are no regulatory restrictions mandating specific operational speed thresholds, speeds of the shuttles could go up to 30–40 km/h, with an average of 20–25 km/h inside cities

As a complement to the public transport network, this service would be priced comparably and integrated with the system both in the mapping and the payment

SHOW sites: *Linköping, Gothenburg, Graz, Salzburg, Carinthia, Monheim, Trikala, Tampere, Turin*

**Scenario 2** Door-to-door delivery of persons and goods

A shared, on-demand, point-to-point service with dynamic routing when or where demand is low, using automated shuttles. This service is detached from a fixed route or primary purpose (e.g. first/last mile). Passengers can be picked up and dropped off in locations of their choosing (DRT), though it may be possible that these points are fixed for efficiency purposes, and may require a short walk. Waiting times do not exceed 10 min, and walking time to the nearest pick up point does not exceed 5 min. While these shuttles would have the same speed as those in scenario 1, the nature of the service (not a fixed route) allows for faster average speeds, thus the service could operate at average speeds in cities of up to 30 km/h (when no regulatory restrictions are in place)

Because of the additional flexibility of this service, its price is higher compared to the fixed route automated shuttles and public transport

SHOW sites: *Les Mureaux (Ariane group private site)*

**Scenario 3** Mass transit AV services

This service is replacing existing PT (mostly bus) lines with a shared automated shuttle or bus. The route of the automated buses runs **between the city centre and points of interest for citizens**, and the bus runs on a fixed route with fixed stops. Passengers wait at the predefined bus stations and are informed for the bus arrival time via their mobile application, if available. The bus stations are also equipped with the bus schedule. The bus follows the fixed route and stops at each station where passengers are detected. The experience in the automated bus is comparable to current public transport with other/ conventional vehicles, with similar comfort and privacy levels. Flexibility is not so high considering the service runs on fixed schedules and routes. With improved efficiency, the frequency is high, with a bus every 4–6 min, and average speeds would be a bit higher than those of shuttles, reaching up to 40–45 km/h (when no regulatory restrictions are in place)

SHOW sites: *Madrid, Brno*

**Scenario 4** Shared Robotaxis

Robotaxis are a point-to-point shared service that operate like regular taxis. Journey reservation is on-demand and the user is picked up at their location. This service is available for private use and sequential sharing (sharing of the vehicle, not trips), and can be also booked in some cases via a mobile application. The service is not part of the public transport network and serves different areas (high and low demand, urban, suburban), often connecting dense urban city centres to residential areas or to any places of interest. The route is dynamic and not fixed, intermediate stops for picking up other passengers are possible if accepted by the passenger and depending the specific service policy. Total extra travel time should be no more than 10 min and average waiting time also around 10 min. The service is by nature flexible, as passengers are dropped off at their selected destination. Due to the flexible routing and small size of the vehicles compared to shuttles and buses, speeds are relatively high, and could go up to 80 km/h (when no regulatory restrictions are in place)

Considering the additional comfort, privacy, and flexibility factors, prices for such a service are expected to be significantly higher than automated shuttles and regular public transport services

SHOW sites: *Madrid, Graz, Trikala, Brno, Karlsruhe*

**Table 2** Questions of the modified Delphi

| Number | Impact | Question | Answer slider scale |
|---|---|---|---|
| 1 | Public transport accessibility | How much do you expect **Public Transport accessibility** to change (in terms of number of transport services in reach within a given time buffer) | *–100%*: high decrease, *0%*: no effect, *100%*: high increase |
| 2 | Public space consumption | How do you expect **public space** consumption to be affected? | *–100%*: less space used, meaning more space freed up and available, *0%*: no variation, *100%*: higher levels of space usage |
| 3 | Public transport equity | How much do you expect **Public Transport equity** and inclusion to change? (regarding physical accessibility, social integration, cohesion, and equality) | *–100%*: high decrease, *0%*: no effect, *100%*: high increase |
| 4 | Housing prices | How much do you expect **housing prices** in the area to vary? | *–100%*: high decrease, *0%*: no effect, *100%*: high increase |
| 5 | Perceived safety | How much do you expect users' **perceived safety** in the area to vary? | *–100%*: high decrease, *0%*: no effect, *100%*: high increase |
| 6 | Job creation/destruction | What direct effects do you expect to take place in terms of **job creation/destruction**? | *–100%*: high job decrease, *0%*: no effect, *100%*: high job increase |
| 7 | Job modification/reskilling | What direct effects do you expect to take place in terms of **job modification/re-skilling**? | *–100%*: heavy job simplification, *0%*: no effect, *100%*: strong job re-skill increase |

and Barela a standard deviation smaller than 1,5 reaches consensus (Christie and Barela 2005).

At the same time, each scenario is assessed against three phases of implementation from piloting (phase 1) to full autonomy[4] (phase 3) with a simultaneous evolving maturity of the technology, penetration rate, and regulatory framework as described below:

- **Phase 1** corresponds to a situation where the service is introduced for testing, to see if the route, the service, and the model are a good fit. This phase also serves to validate whether the innovative service meets the needs, expectations, and daily journeys of citizens. Vehicles and their technology are at an early stage (SAE L3 to L4), travelling at low speeds and requiring an on-board driver for safety and emergency reasons, as automation and infrastructure are not fully prepared.

---

[4] "Full autonomy" refers to SAE Level 5 of automation by the J3016 standard, that defines six levels of automation, from 0 to 5, source: 2024 SAE International (SAE International).

Penetration rate and technology maturity as well. Effects on cities and public are expected to be low, however new needs in employment or infrastructure may be revealed. This phase serves as a baseline for comparison and corresponds to the SHOW pilot cases interviewed.

- **Phase 2** represents a higher penetration rate than phase 1, where the service has been tested and acceptance is clear, as evidenced by a higher number of users, clearly defined routes, and citizens' awareness of the service, both in terms of use and coexistence. Accordingly, the infrastructure and vehicles both are characterised by a higher readiness, as is the automation of the service (up to SAE L4), allowing for less dedicated human control, although still needed.

- **Phase 3** represents the full deployment of innovative services and penetration rate, through consolidation and even replication. The benefits are clear, and citizens massively use to benefit from the new mobility solutions. Similarly, infrastructure and vehicles are progressively upgraded (towards SAE level 5 of automation), eliminating the need for on-board human support.

In parallel to the modified Delphi study, interviews with the pilot sites of Monheim (Scenario 1) and Les Mureaux (Scenario 1 and Scenario 2 as they provided two different services) were conducted to verify the assumptions from the literature and to unveil potential unique issues of their pilot sites. These interviews aimed to understand the needs of the sites in terms of employment and skills requirements, as well as the real-life experiences in the practical case of a pilot implementation as in SHOW (Phase 1).

The external expert interviews (n = 4) were conducted online with experts in urban mobility and CCAV with long experience in CCAV testing and European projects. These interviews lasted about one hour each and covered the topics described above. In particular, they were asked about their views regarding the impact of CCAV on society and employment, prompt also to provide future recommendations.

## 4 Findings

Table 3 shows the questions, the average responses, and the standard deviation for the two rounds, across each scenario and throughout the three (3) phases. There were 78 respondents in the first round and 40 in the second. In most cases where consensus was reached in the second round (SD < 1, 5), this is highlighted in bold.

In general, Delphi participants reached an agreement with a standard deviation of less than 1, 5 in the second round. There tends to be less agreement regarding S4 shared robotaxis, likely due to its limited deployment in Europe compared to the other scenarios, with participants generally more familiar with the latter. Similarly, there is less agreement in phase 3 than in the other 2 phases, as it is a more distant future situation and there is more uncertainty among the participants.

The results of the modified Delphi show an increase of **accessibility to public transport** on average with higher penetration rates of the service in S1, S2 and S3,

**Table 3** Results of the modified Delphi

| | Scenario 1 (S1) Driverless shuttle for first/last mile | | | | Scenario 2 (S2) Door-to-door delivery of persons and goods | | | | Scenario 3 (S3) Mass transit AV services | | | | Scenario 4 (S4) Shared Robotaxis | | | |
|---|---|---|---|---|---|---|---|---|---|---|---|---|---|---|---|---|
| | Round 1 | | Round 2 | | Round 1 | | Round 2 | | Round 1 | | Round 2 | | Round 1 | | Round 2 | |
| | Avg. | SD | Avg. | SD | Avg. | SD | Avg. | SD | Avg. | SD | Avg. | SD | Avg. | SD | Avg. | SD |
| *Public transport accessibility* | | | | | | | | | | | | | | | | |
| Phase 1 | 12,6 | 1,6 | 6,4 | 0,7 | 12,5 | 1,1 | 6,7 | 0,5 | 9,4 | 1,7 | 8,5 | 0,8 | 6,0 | 1,7 | 2,9 | 0,5 |
| Phase 2 | 28,6 | 1,6 | 20,9 | 0,8 | 29,2 | 1,7 | 16,7 | 0,5 | 26,5 | 1,5 | 18,6 | 1,3 | 2,1 | 1,7 | 4,3 | 1,4 |
| Phase 3 | 44,7 | 2,3 | 36,1 | 1,3 | 58,3 | 2,5 | 26,7 | 1,0 | 37,4 | 2,7 | 30,0 | 1,9 | -2,9 | 3,6 | 0,0 | 3,3 |
| *Public space consumption* | | | | | | | | | | | | | | | | |
| Phase 1 | 3,8 | 1,3 | 1,8 | 0,4 | 4,0 | 1,2 | 2,5 | 0,5 | -7,7 | 1,6 | -1,8 | 0,9 | 2,2 | 1,0 | 2,9 | 0,5 |
| Phase 2 | 8,3 | 2,2 | 4,0 | 0,9 | 1,8 | 2,3 | 2,5 | 0,5 | -2,1 | 2,2 | -1,0 | 0,7 | 3,6 | 2,0 | 5,7 | 1,6 |
| Phase 3 | 13,4 | 3,6 | 5,7 | 1,4 | -4,2 | 3,7 | 0,0 | 0,8 | 4,4 | 3,6 | 0,0 | 1,1 | -7,3 | 3,5 | -5,7 | 3,7 |
| *Public transport equity* | | | | | | | | | | | | | | | | |
| Phase 1 | 7,8 | 1,7 | 5,7 | 0,7 | 9,2 | 1,2 | 6,7 | 0,5 | 8,1 | 1,4 | 6,7 | 0,5 | 1,8 | 1,2 | 5,0 | 0,9 |
| Phase 2 | 23,5 | 1,6 | 16,8 | 0,9 | 19,2 | 2,8 | 10,0 | 1,1 | 16,8 | 1,6 | 15,7 | 1,2 | -2,1 | 2,5 | 7,5 | 1,2 |
| Phase 3 | 35,5 | 2,0 | 28,3 | 1,3 | 36,7 | 4,1 | 13,3 | 1,9 | 28,9 | 2,7 | 26,4 | 2,0 | -3,6 | 5,0 | 11,3 | 3,2 |
| *Housing prices* | | | | | | | | | | | | | | | | |
| Phase 1 | 2,8 | 1,2 | 1,3 | 0,3 | 3,3 | 0,5 | 0,0 | 0,0 | 4,2 | 1,4 | 2,9 | 0,6 | 1,7 | 0,4 | 2,0 | 0,4 |
| Phase 2 | 9,2 | 1,0 | 7,1 | 0,7 | 15,7 | 1,4 | 6,7 | 0,5 | 15,0 | 1,4 | 9,3 | 0,9 | 13,3 | 1,5 | 5,0 | 0,8 |
| Phase 3 | 16,4 | 1,6 | 12,2 | 1,0 | 19,1 | 1,7 | 10,0 | 0,6 | 24,4 | 2,6 | 17,1 | 1,3 | 25,0 | 2,2 | 10,0 | 1,3 |

(continued)

**Table 3** (continued)

| | Scenario 1 (S1) Driverless shuttle for first/last mile | | | | Scenario 2 (S2) Door-to-door delivery of persons and goods | | | | Scenario 3 (S3) Mass transit AV services | | | | Scenario 4 (S4) Shared Robotaxis | | | |
|---|---|---|---|---|---|---|---|---|---|---|---|---|---|---|---|---|
| | Round 1 | | Round 2 | | Round 1 | | Round 2 | | Round 1 | | Round 2 | | Round 1 | | Round 2 | |
| | Avg. | SD | Avg. | SD | Avg. | SD | Avg. | SD | Avg. | SD | Avg. | SD | Avg. | SD | Avg. | SD |
| *Perceived safety* | | | | | | | | | | | | | | | | |
| Phase 1 | −0,5 | 1,6 | **−1,9** | **0,7** | 0,0 | 1,1 | **0,0** | **0,7** | 0,6 | 2,7 | **−2,7** | **0,9** | −5,5 | 1,8 | **−6,3** | **0,7** |
| Phase 2 | 12,8 | 1,9 | **5,5** | **0,9** | 11,1 | 1,8 | **5,0** | **0,5** | 10,6 | 2,0 | **6,9** | **1,0** | 5,0 | 1,9 | **2,9** | **1,5** |
| Phase 3 | 23,6 | 2,4 | **12,7** | **1,2** | 14,5 | 2,2 | **8,3** | **0,8** | 21,5 | 2,4 | 16,4 | 1,7 | 16,4 | 3,8 | 15,0 | 3,6 |
| *Job creation/destruction* | | | | | | | | | | | | | | | | |
| Phase 1 | 4,9 | 0,8 | **3,2** | **0,5** | 6,7 | 0,7 | **4,0** | **0,5** | 12,1 | 3,0 | **5,0** | **0,5** | −2,5 | 1,0 | **0,0** | **0,8** |
| Phase 2 | 3,0 | 1,2 | **3,3** | **0,8** | 6,4 | 1,3 | **2,0** | **1,3** | 3,3 | 1,8 | **2,7** | **1,0** | −10,0 | 1,4 | −1,3 | 1,7 |
| Phase 3 | 0,7 | 2,1 | **0,0** | **1,1** | 0,0 | 2,6 | −7,5 | 1,9 | 0,0 | 2,2 | −2,7 | 1,8 | −19,3 | 2,6 | −11,3 | 2,9 |
| *Job modification/re-skilling* | | | | | | | | | | | | | | | | |
| Phase 1 | 9,5 | 1,3 | **6,2** | **0,5** | 8,6 | 0,7 | **6,0** | **0,5** | 12,9 | 1,8 | **6,4** | **0,7** | 7,1 | 0,8 | **5,0** | **0,5** |
| Phase 2 | 17,6 | 2,2 | **15,9** | **0,9** | 22,7 | 0,9 | **13,3** | **0,5** | 22,8 | 1,8 | **17,5** | **0,9** | 13,8 | 1,4 | **16,3** | **1,1** |
| Phase 3 | 28,0 | 3,3 | **24,1** | **1,5** | 40,8 | 1,6 | **20,0** | **1,3** | 31,0 | 2,4 | **24,3** | **1,3** | 22,1 | 2,4 | 31,3 | 1,9 |

while shared robotaxis (S4) does not reach a consensus. This may be a response to the uncertainty of its potential customers, as some experts argue that the prices of the service are likely to be higher than the others, which could affect affordability and, therefore, accessibility. This aligns with the results of the interviews to the pilots (Monheim and les Mureaux) for the long-term vision, which foresees an increase in accessibility in the future. The pilot in Monheim has implemented automated bus lines connecting the old town to a bus terminal and has improved accessibility for groups such as the elderly and children, as narrow streets not accessible by regular buses now have a public transport service. The shuttles also travel to Monheim's health campus, serving workers and patients, many of whom being elderly.

**Equity** results show a clear improvement of equity for scenarios 1 (first/last mile) and 3 (mass transit), and less improvement for scenarios 2 (door-to-door) and 4 (shared robotaxis), probably due to the cost of service for the customers. At the private site of Les Mureaux, the passengers are employees of the Ariane Group and the pilot has received a positive response from users, although it usually replaces walking on the site. The open road pilot (also in Les Mureaux), which covers the last mile from the train station to the facility, has had a very positive response from users, who complained when it stopped as it left a gap in the network. However, the shuttle is sometimes not as effective when using the ramp, which is not fully integrated and has a negative impact on the use of the service by people with physical disabilities.

Most European cities are in the process of removing space for cars and parking, and shared AVs can help free up **public space**, which is likely to be more available with a high penetration rate for S2 door-to-door delivery of people and goods, and S4 shared robotaxis, is not the case for regular lines services (S1). This result has different interpretations; when asked to the pilot representatives, the prediction is that when the service is largely extended, usage of private cars shift to automated public transport could free up space, however, some experts argue that the shift could be small and so would be the impact on space. Pilots are currently using existing infrastructure without the need to build new street infrastructure, other than reserving some space for their operations by removing parking spaces. Sometimes the nature of the site does not allow for any changes (pilot in Les Mureaux) and just some blocks in a parking lot are enough to develop the pilot. In the long term, it is expected that fewer private cars will be needed, and public space will be gained (second to third phase in the Delphi).

The impact on **house prices** is expected to have an increase in all scenarios, confirming previous studies which argue that better public transport including automated services increases land prices and house prices respectively.

**Perceived safety** by users, is one of the key elements to scale up automated services according to the pilots and experts interviewed and experts agree that is expected to grow as penetration rate increases and, mostly as technologies improve. This will depend largely on the technology improvements, but, also, on the general acceptance of riding an automated shuttle without human action, which is likely to increase in the future according to the average answers.

In the early stages of AV deployment, **employment** opportunities are expected to increase, as regular services will continue as usual, and new automated services will need to hire or train new profiles of employees. This will come along with a clear trend towards re-skilling across all services, which will increase as penetration increases. However, as penetration rates increase, the overall employment landscape is expected to stabilise, reflecting the current scenario. In particular, S4 shared robotaxis do not reach a clear consensus and show a decrease in employment, which can be explained by the easy scalability of the service once the technology and regulations allow it to be fully automated and monitored with minimal human intervention. However, jobs will change significantly as there is a clear trend towards re-skilling across all services, which will increase as penetration increases. The SHOW pilots have also had an impact on employment. Although this has been limited due to the scale of the operations, we can already provide some insight into what will be needed in the future. In the case of Monheim, about 50 people were trained to become on-board operators, about 30 of which remained in the new role and were needed to supervise the shuttle. Around a third of drivers did not want to continue this type of operation and returned to regular driving. The main reasons for this were conflicts with other road users (e.g., angry drivers not used to have automated vehicles on the road), new challenging or more demanding tasks arisen in their new role and the fact that they had to stand in the vehicle for a long time. This differs from other pilots where safety drivers found the task unchallenging and boring, so we can see that the context and nature of the service matters in the experience of on-board operators.

In the case of Les Mureaux, the drivers trained for the new role as on-board operators and remote operators were preselected and willing to participate, and the response from the employees was mostly positive. They had to undertake one (on-board) or two (remote) trainings, and those that remained until the end faced some kind of difficulties when going back to their original role of driving a manual bus. Some drivers did not want to go back to the previous roles and, instead, they pursued the role of a trainer for automated driving supervision. This transition resulted in an upgrade from their previous tasks.

These new tasks, which are usually more demanding and require training, have been compensated in salary or bonuses in both cases. And when it comes to hiring new people, they have different profiles than any type of employee hired before, which focuses on automated driving supervision skills.

The interviews with external experts corroborated the findings regarding the impact on jobs from the modified Delphi and added to the challenges faced and potential fields of study. They highlighted that it is too early to say what impact automation can have on employment, but it will be directly linked to the business model, as seen in the Delphi results. The regulatory framework is also key to the impact on society, and the U.S. and Europe are taking different approaches to this, with the U.S. being able to roll out services and test the technology faster, but with more uncertainty about the societal impact. Europe has had extensive experience of the consequences of automation that goes hand in hand with digitalisation, and according to experts, one of the main drivers of automation is the reduction of labour costs. An illustrative example of the impact on jobs is evident in the implementation of the automated metro line in Turin, which has not resulted in fewer employees but rather in a shift towards different profiles, such as the replacement of drivers with engineers. This relates to the challenge faced by public authorities and PTOs in recruiting and retaining talent, especially IT professionals due to the competitiveness and high demand for such profiles. This is added to the potential challenge posed by automation, which could potentially move digital jobs, such as IT support abroad, with implications for both employment and the economy.

# 5 Conclusion

In summary, our study shows the potential impacts of CCAV deployment on accessibility, equity, and employment dynamics. Through a modified Delphi method and interviews, we gained insights into the evolving transportation landscape.

Our findings suggest that while CCAV holds promise for enhancing accessibility to public transport, challenges persist, particularly regarding affordability and inclusivity, notably with regard to the deployment of shared robotaxi services. Moreover, the transition to automated services presents both opportunities and challenges for employment, emphasising the importance of proactive measures in workforce planning.

However, it is important to acknowledge the limitations of the study. The reduced participation in the second round of the Delphi study (40 participants compared to 78 in the first round) may have influenced the consensus reached. In addition, the scope of the study was limited to certain pilot regions and operations, which may affect the generalisability of our findings, for example, regarding the limited examples of shared robotaxis in Europe.

We recommend future research to further investigate the societal implications of different services of CCAV deployment to ensure a smooth transition towards a more sustainable and equitable transportation ecosystem.

**Acknowledgements** The research presented is a part of the SHOW project that has received funding from European Union's Horizon 2020 research and innovation programme under grant agreement no. 875530.

# References

Alonso Raposo M, Grosso M, Després J, Fernandez Macias E, Galassi M, Krasenbrink A, Krause J, Levati L, Mourtzouchou A, Saveyn B, Thiel C, Ciuffo B (2018) An analysis of possible socio-economic effects of a Cooperative, Connected and Automated Mobility (CCAM) in Europe. EUR 29226 EN, Publications Office of the European Union, Luxembourg. https://doi.org/10.2760/777 (online), https://doi.org/10.2760/007773 (print)

Beiderbeck D, Frevel N, von der Gracht HA, Schmidt SL, Schweitzer VM (2021) Preparing, conducting, and analyzing Delphi surveys: cross-disciplinary practices, new directions, and advancements. MethodsX. https://doi.org/10.1016/j.mex.2021.101401

Brown P-M (2014) Autonomous vehicles: a thought leadership review of how the UK can achieve a fully autonomous future. The Institution of Engineering and Technology, Stevenage

Bruzzone F, Cavallaro F, Nocera S (2022) The definition of equity in transport. In: AIIT 3rd international conference on transport infrastructure and systems (TIS ROMA 2022). Rome, Italy. https://doi.org/10.1016/j.trpro.2023.02.193

Christie CA, Barela E (2005) The Delphi technique as a method for increasing inclusion in the evaluation process. Can J Program Eval 20(1):105–122. https://doi.org/10.3138/cjpe.020.005

Ciuffo B, Raposo MA (2019) The future of road transport: Implications of automated, connected, low-carbon and shared mobility. In: European commission, joint research centre

Cohen T, Cavoli C (2019) Automated vehicles: exploring possible consequences of government (non)intervention for congestion and accessibility. Transp Rev 39(1):129–151. https://doi.org/10.1080/01441647.2018.1524401

Custer RL, Scarcella JA, Stewart BR (1999) The modified delphi technique—a rotational modification. J Vocat Tech Educ 15(2):50–58. https://doi.org/10.21061/jcte.v15i2.702

Dalkey N, Helmer O (1963) An experimental application of the DELPHI method to the use of experts. Manag Sci 9(3):458–467. https://doi.org/10.1287/mnsc.9.3.458

Eby DW, Molnar LJ, Zhang L, Louis RMS, Zanier N, Kostyniuk LP, Stanciu S (2016) Use, perceptions, and benefits of automotive technologies among aging drivers. Inj Epidemiol 3:1–20. https://doi.org/10.1186/s40621-016-0093-4

European Commission (2021) Directorate-General for Mobility and Transport. Study on the social dimension of the transition to automation and digitalisation in transport, focusing on the labour force: final report, Publications Office. https://doi.org/10.2832/95224

Gelauff G, Ossokin I, Teulings C (2017) Spatial effects of automated driving: dispersion, concentration or both? https://doi.org/10.13140/RG.2.2.32766.48965

Giannarou L, Zervas E (2014) Using Delphi technique to build consensus in practice. Int J Bus Sci Appl Manag 9(2):65–82

Harper C, Hendrickson C, Mangones S, Samaras C (2016) Estimating potential increases in travel with autonomous vehicles for the non-driving, elderly and people with travel-restrictive medical conditions. Transp Res Part c: Emerg Technol 72:1–9. https://doi.org/10.1016/j.trc.2016.09.003

ITF (2023) Making automated vehicles work for better transport services: regulating for impact. international transport forum policy papers, No. 115. OECD Publishing, Paris. https://doi.org/10.1787/2ea70307-en

Koul S, Eydgahi A (2020) The Impact of Social Influence, technophobia, and perceived safety on autonomous vehicle technology adoption. Period Polytech Transp Eng 48(2):133–142. https://doi.org/10.3311/PPtr.11332

Landeta J (2006) Current validity of the Delphi method in social sciences. Technol Forecast Soc Change 73(5):467–482. https://doi.org/10.1016/j.techfore.2005.09.002

Litman T (2024) Evaluating public transit benefits and costs. Victoria Transport Policy Institute

Milakis D, Arem B, Wee B (2017) Policy and society related implications of automated driving: a review of literature and directions for future research. J Intell Transp Syst 21(4):324–348. https://doi.org/10.1080/15472450.2017.1291351

Moreno C (2021) Introducing the "15-Minute City": sustainability, resilience and place identity in future post-pandemic cities. Smart Cities 4(1):93–111. https://doi.org/10.3390/smartcities4010006

Okamoto T (2019) Japan faces urgent need to develop autonomous transport due to graying society and driver shortage. The Japan Times. www.japantimes.co.jp/news/2019/09/22/national/japan-faces-urgent-need-develop-autonomous-transportation-system-due-graying-society-shortage-drivers

SAFE Center (2023). Increasing mobility and access with autonomous vehicles. Issue Brief. https://secureenergy.org/increasing-mobility-access-avs/

SAE International: SAE Levels of Driving Automation Refines for Clarity (n.d.) Accessed from https://www.sae.org/standards/content/j3016_201806/

Scanlon JM, Kusano KD, Engström J, Victor T (2022) Collision avoidance effectiveness of an automated driving system using a human driver behavior reference model in reconstructed fatal collisions. Waymo, LLC

Schéma directeur de la Région Île-de-France (SDRIF-E) (2040) Un nouvel équilibre. Webpage, 2023. Accessed 16 April 2024. https://www.institutparisregion.fr/planification/ile-de-france-2040/sdrif-e-2040/

Shariff A, Bonnefon J-F, Rahwan I (2017) Psychological roadblocks to the adoption of self-driving vehicles. Nat Hum Behav. https://doi.org/10.1038/s41562-017-0202-6

Skillful (2019) D 5.4: Best practices application guidelines and policy recommendations, deliverable of the horizon 2020 skillful project, Grant Agreement No 723989 https://skillfulproject.eu/ (upcoming online)

WeTransform (2020) D3.1: Report of actions and initiatives related to transport automation and other transitions, Deliverable of the Horizon 2020 WeTransform project Grant Agreemtn No 101006900, https://wetransform-project.eu/site/asets/files/1724/wet_d3_1_220415_report_of_actions_and_initiatives_related_to_transport_automation_and_other_transitions.pdf

Whitmore A (2022) Integrating shared autonomous mobility into the U.S. transportation system: an equity, economic, ethical, and environmental assessment. Carnegie Mellon University, ProQuest Dissertations Publishing. https://doi.org/10.1184/R1/21861666.v1

Xu Z, Zhang K, Min H, Zhao X, Liu P (2018) What drives people to accept automated vehicles? findings from a field experiment. Transp Res Part c: Emerg Technol 95:320–334. https://doi.org/10.1016/j.trc.2018.07.024

Zhang Q, Wallbridge CD, Jones DM, Morgan PL (2024) Public perception of autonomous vehicle capability determines judgment of blame and trust in road traffic accidents. Transp Res Part a: Policy Pract 179:103887. https://doi.org/10.1016/j.tra.2023.103887

# Stakeholders' Engagement in Shared Automated Mobility: A Comparative Review of Three SHOW Approaches

Delphine Grandsart, Kathryn Bulanowski, Henriette Cornet,
Fatima-Zahra Debbaghi, Matina Loukea, Maria Gkemou, Petra Schoiswohl,
and Walter Prutej

**Abstract** In this paper, we explore the importance of citizen and stakeholder engagement in the development of new mobility services, and how such aspects have been integrated and applied in the EU funded project SHOW (Horizon 2020 GA No. 875530). First, we provide a broad overview of how we engaged end-users and stakeholders in the project's different pilot sites. Next, we zoom in on three engagement mechanisms—the Ideathon in Carinthia (Austria), the Hackathon in Thessaloniki (Greece) and the MAMCA (Multi-Actor Multi-Criteria Analysis) workshop in

D. Grandsart (✉) · K. Bulanowski
European Passengers' Federation (EPF), Kortrijksesteenweg 304, 9000 Gent, Belgium
e-mail: delphine.grandsart@epf.eu

K. Bulanowski
e-mail: kathryn.bulanowski@epf.eu

H. Cornet
University of San Francisco (USF), Autonomous Vehicles & the City, 2130 Fulton St., San Francisco, CA 94117, USA
e-mail: hcornet@usfca.edu

F.-Z. Debbaghi
Electromobility Research Centre, Vrije Universiteit Brussel (VUB), Pleinlaan 2, 1050 Brussels, Belgium
e-mail: Fatima-zahra.debbaghi@vub.be

M. Loukea · M. Gkemou
Centre for Research and Technology Hellas (CERTH), Hellenic Institute of Transport (HIT), 34 Ethnarchou Makariou St., 16341 Athens, Greece
e-mail: mloukea@certh.gr

M. Gkemou
e-mail: mgemou@certh.gr

P. Schoiswohl · W. Prutej
Pdcp GmbH, Smart Urban Regional Austria Alps Adriatic (SURAAA), Hauptstraße 204, 9210 Pörtschach, Austria
e-mail: petra.schoiswohl@suraaa.at

W. Prutej
e-mail: walter.prutej@suraaa.at

© The Author(s) 2025
H. Cornet and M. Gkemou (eds.), *Shared Mobility Revolution*, Lecture Notes in Mobility, https://doi.org/10.1007/978-3-031-71793-2_10

Tampere (Finland)—, presenting a variety of participative approaches for designing and evaluating new automated mobility services. The chapter offers a comparative analysis of these three approaches, highlighting for each case the opportunities and challenges. We show that stakeholder engagement activities efficiently generate ideas and validate solutions at a local level enriching the innovation process with novel perspectives, yet resource allocation and participant diversity pose challenges.

**Keywords** User engagement · Stakeholder participation · Ideathon · Hackathon · Innovation process · Automated mobility · MAMCA

# 1 Introduction

Connected, cooperative and automated mobility (CCAM) has the potential to contribute to a safer, more affordable, more inclusive and sustainable mobility, thus bringing great benefits to both citizens and society. In the development of new mobility paradigms such as CCAM, the engagement of both citizens and stakeholders is useful and necessary, to make sure that services are designed to meet their needs and requirements and to increase the positive impacts on society (Grandsart et al. 2023).

Different theoretical models exist that aim to categorize different forms of citizen and stakeholders' engagement. For example, in Arnstein's "ladder of participation" (Arnstein 1969), each step of the metaphorical ladder corresponds to a certain degree of citizen agency, control and power to influence the decision-making process. Informing and consulting are activities that are situated on the low end of the participation ladder, whereas concepts such as co-creation, co-design, living labs, etc. are characterized by a higher level of participation, empowering people and sometimes even placing the final decision-making in the hands of the public. Within the SHOW project,[1] a wide variety of citizen and stakeholder engagement activities have been deployed, ranging from a low level of participation (informing people about the automated vehicle services, a-priori and a-posteriori surveys as described in SHOW 2020) to a higher level of participation (such as Ideathons, Hackathons, focus groups, workshops).

In this chapter, three engagement mechanisms are discussed in more detail, in terms of objectives, addressed stakeholders, methodology and effectiveness:

- the SHOW Ideathon in Carinthia (Austria) (Chap. 3)
- the SHOW Hackathon in Thessaloniki (Greece) (Chap. 4)
- the SHOW MAMCA workshop in Tampere (Finland) (Chap. 5).

---

[1] SHOW (SHared automation Operating models for Worldwide adoption) has received funding from the European Union's Horizon 2020 research and innovation programme under grant agreement No 875530. More information https://www.show-project.eu/.

A comparative review of these three types of activities is provided, highlighting for each case the opportunities and challenges, as well as recommendations (Chap. 6).

Together, the approaches considered in this paper represent a triangulation of engagement mechanisms, considering a variety of aims covering the innovation process (Verworn and Herstatt 2002) from *idea generation* through passengers' vision of future services (Ideathon), over new or added value services *development* by researchers and developers (Hackathon), to *validation* of Automated Vehicles (AV) services by key stakeholders of the CCAM value chain (MAMCA workshop) namely users, local authorities, transport operators, Original Equipment Manufacturers (OEMs), infrastructure operators (if any), service providers (related fleet management, security services or add-on services), public interest groups, and researchers.

While several Ideathons, Hackathons and MAMCA workshops were conducted within the SHOW project, one case per each is discussed in this chapter. Furthermore, it is worth noting that, in all three selected cases, the participants involved had not experienced, directly or indirectly, the piloted services. An example of an engagement activity in SHOW that focused on involving users in testing the services (the 'Supertesters' approach as applied in Austria) is described in the first chapter of part 3 of this contributed volume.

## 2   Theoretical Background and Case Studies Selection

In the realm of developing and implementing driverless mobility, it is imperative for industry stakeholders and policymakers to actively engage lay citizens in order to solicit their inputs, address their concerns, and understand their aspirations. For instance, extensive engagement efforts have occurred with the organization of Citizens' Dialogues in 15 cities across eight countries involving 945 citizens. These Dialogues aimed to explore CCAM initiatives globally, spanning various service contexts, and conducted cross-cultural analyses to discern differences in perception based on geographical location (Chng et al. 2021).

The findings from the Dialogues indicated that the public generally holds positive attitudes towards CCAM, albeit with typical concerns regarding safety and vehicle capabilities (Chng et al. 2021).

However, these studies do not contribute to finding solutions to address these concerns or support local decision-making processes. Therefore, in this chapter, we offer some inspiration on how to extend beyond mere user acceptance and adoption intention of CCAM by actively involving a variety of stakeholders throughout the innovation process, as exemplified in the SHOW project.

We refer to the innovation process as described by Leven and Kersten (2021) with the stages of idea generation, development, validation, and commercialisation, in which we exclude the commercialisation phase because the project works on a pre-commercial basis.

By actively engaging a variety of stakeholders, our aim is to maximize their contributions and ensure that the solutions developed align closely with a broad spectrum of needs and expectations. Moreover, this approach is intended to uncover aspects that may not have been initially considered by experts, thereby enriching the innovation process with novel insights and perspectives.

In this paper, we are interested in three engagement mechanisms applied in SHOW that had similarities in specific key attributes (structured in-person workshop settings with selected participants embedded in a local context) but varied in their aim and phase in the innovation process:

i.   Idea generation by future potential users
ii.  Development by researchers and developers
iii. Assessment/validation of services to support decision-making by a mix of stakeholders.

For (i), the method of Ideathon has been selected. An Ideathon brings people from various backgrounds together to create innovative ideas within a certain period, which can range from a few hours to several days, usually focusing on a specific topic or challenge. Individuals or teams brainstorm, discuss and refine their ideas, before presenting them to a jury or panel of experts. Participants are encouraged to think outside of the box in an environment that stimulates creativity. Ideathons provide a platform for networking, skill building and the development of viable concepts that can be further developed into effective solutions or projects (Sakiyama et al. 2020).

For (ii), a Hackathon has been organized. A Hackathon is an intensive collaborative event that brings together professionals, developers, designers and subject matter experts to solve specific problems or create innovative solutions within a limited timeframe—often spanning a day or a weekend (Hackathon.com 2021). Participants typically form teams and work on projects related to software development, hardware, data analysis, or other technological challenges, aiming to produce tangible outcomes, such as prototypes, proofs of concept or functional solutions. The term "hack" in Hackathon is used in a positive sense, indicating a creative and dynamic approach to problem-solving rather than any malicious activity (Jones et al. 2015). Hackathons are widely employed in various fields. In the context of mobility research, Hackathons can be utilized to address specific challenges related to mobility, traffic management, or the integration of emerging technologies such as CCAM.

For (iii), the multi-actor, multi-criteria analysis method (MAMCA) has been tested, in which stakeholders are brought together in a workshop to evaluate different alternatives across multiple criteria (Macharis et al. 2009). The MAMCA method was developed at the Vrije Universiteit Brussel (VUB) with the goal of incorporating the opinions of stakeholders in a multi-criteria decision making approach (Macharis 2005), and can be applied to evaluate transport infrastructure projects and policies (Cornet et al. 2018; Sun et al. 2015; Lebeau et al. 2018) and also different automation services (Feys et al. 2020). In a workshop setting, stakeholders are assigned to breakout groups, in which they first provide weights to criteria reflecting their importance, and then evaluate the performance of different scenarios in terms of each criterion. Aggregated scores are calculated using the criteria weights and the evaluation scores, and are later visualized and presented for discussion in a plenary setting with all stakeholders. An online tool (available on https://www.mamca.vub.be) was developed to facilitate the application of the methodology (Huang et al. 2020).

All three engagement mechanisms included a phase of preparation with a promotional campaign on social media and through newsletters, emails and sometimes (as for (i)) even direct phone calls, to engage potential participants. Catering was offered to the participants as well as several incentives such as goodie bags (in the case of (i)), shopping vouchers (i) and a prize for the winning teams. In the case of (i), the prize was an additional EUR 500, and in (ii), an invitation to international events. At the start of the day, all three methods included a preliminary informational/educational session where the participants were made familiar with the process for the day(s) and with the topic of automated shared mobility (Table 1).

Various ways of evaluating the effectiveness of such engagement have been extensively discussed in Rowe and Frewer (2005). In this chapter, the following criteria have been applied by the authors to assess effectiveness of the selected mechanisms:

- **Quality of insights**: how innovative, creative, original are the outcomes and discussions, to which extent do they successfully address the stated challenges or objectives, are they feasible and practical to implement in real life, and would it be possible to adapt or scale up?
- **Collaboration and team dynamics**: did participants positively collaborate and communicate, were they able to acquire new skills, insights or knowledge, how were the overall group dynamics?
- **Community engagement and post-event impact**: did the event attract interest from the general public, media and (other) stakeholders, did it lead to follow-up projects or collaborations, or policy changes?

**Table 1** Case studies overview

| Method | Stage of the innovation process | Participants | Aim | Duration |
|---|---|---|---|---|
| Ideathon | Idea generation | N = 52, including 29 non-experts | Brainstorm on predefined challenges. Generate ideas to solve challenges | One day |
| Hackathon | Development | N = 20, including students, software developers and young researchers | Develop services related to the use of CCAM, focusing on three topics selected from a previous Ideathon | Three days |
| MAMCA | Assessment/ Validation | N = 24, including users, local authorities, operators, associations and researchers | Evaluate impacts of different automation scenarios | Less than half a day |

- **Challenges**: how challenging was the engagement method on organizational matters?

The following sections present the results for each engagement mechanism.

## 3  Ideathon (Carinthia, Austria)

On 5 July 2023, SURAAA (Smart Urban Region Austria Alps Adriatic) organized an Ideathon in the context of the SHOW project. It was a full-day event (from 8:30 AM until 7:30 PM, followed by an evening program) in Pörtschach, that gathered 52 participants: 29 citizens of different ages and backgrounds, 10 SURAAA employees, 12 external experts and one external process support person from the Carinthian University of Applied Sciences.

The objective of the Ideathon was to brainstorm and develop innovative solutions for the mobility of the future, addressing three questions in specific:

1. How to connect public transport (bus, train) with autonomous, on-demand services (shuttles, vehicles, robotaxis, etc.)?
2. How should automated shuttles be designed so that contactless use of the AV (via an app, with monitors, voice output, sensors, etc.) is possible?
3. How to combine the transport of people and small goods to optimize the use of vehicles (i.e., avoid empty trips)?

In order to gain an understanding of the mobility of the future, an automated shuttle used for real-life deployment in Pörtschach was placed in front of the building throughout the day and could be visited by the participants. An operator answered all open questions.

The day started with registration and introductions, followed by presenting the challenges. Nine teams were formed, diverse in age and gender, each assigned to one of three challenges using color-coded dots.

The work process was divided into two phases. By the end of the morning part, each group should have decided on its final idea. The afternoon was dedicated to further concrete development of the ideas. Finally, the ideas were pitched and evaluated by the jury, which consisted of three experts. The evaluation criteria consisted of the following points: presentation, degree of maturity/feasibility, customer benefit/ market potential and innovation content/originality. There was also an audience vote.

Electronic devices like laptops or tablets were not allowed. The participants were free to decide how they would like to present their idea to the audience. For this, they were provided with a wide array of creative materials—modelling clay, flipcharts, craft materials and other analogue materials—to choose from and use during their work and for presenting their ideas.

There were three distinct challenges to address (see above) and each group presented unique solutions:

**Question 1**: Addressing the challenge of poor bus stop facilities on the countryside, *Team DigiStop* proposed a comprehensive upgrade, envisioning dynamic hubs that offer protection from weather conditions and additional amenities to make public transport more attractive for all citizens. *Team 14Fun* presented the concept of transforming train stations into mobility hubs, integrating voice recognition for interactive booking. Their solution suggested additional nodes based on user preferences and availability. *Team App2Anywhere* showcased an integral Mobility as a Service (MaaS) app, enabling route planning, individual vehicle booking and integrated journeys. The app's modular design accommodated diverse user needs, emphasizing inclusivity and comfort.

**Question 2**: To overcome the hurdle of complex ticketing systems hindering public transport adoption, *Team MobilityFlow* introduced an innovative solution to eliminate the need for physical tickets or mobile phones, making public transport more accessible and convenient for everyone. *Team Tourismus +* envisioned a unique combination of autonomous driving and cultural experiences. Their proposal aimed at offering relaxing journeys to cultural hotspots, emphasizing easy online/app payments and interactive cultural exploration. Introducing data-driven solutions, *Team ShuttleBots* analysed customer travel behaviour and optimized occupancy times to improve public transport efficiency, with a focus on enhancing user comfort and overall system efficiency.

**Question 3**: In Austria, there are many rural areas and numerous parcel delivery companies, each catering to specific households. At times, this results in a significant number of unnecessary routes. *Team SMS* (Smart Mobility Solutions) proposed transforming bus stops into central pick-up points for small parcels, seamlessly integrating public transport and parcel delivery. This innovative approach aimed to achieve cost savings for logistics providers, generate additional income for public transport and reduce environmental impact. *Team Automota* presented an automated shuttle solution equipped with modular logistics boxes, addressing issues of underutilization. Their proposal included on-demand delivery services and innovative add-ons like hot holding and cooling modules. With a vision geared towards optimizing package delivery, *Team Kings of Combination* crafted a strategic approach that harnessed the potential of current transportation resources such as trains, buses, taxis and private cars. Central to their solution was the prioritization of real-time route planning and the establishment of Transmobility Hubs as micro depots, all aimed at fostering a seamless and efficient package handling process.

The effectiveness of the method has been assessed as follows:

**Quality of insights**: The elaborated ideas showed a highly innovative character. During pitching, the groups consistently succeeded in presenting their visions in a creative way, without digital aids. The jury members were positively impressed and selected one winning team per challenge (*App2anywhere*, *Shuttle Bots*, *Kings of Combination*). The ideas have brought new solutions for automated shuttles but have also confirmed existing assumptions and expectations, for instance regarding mixed use of shuttles for passengers and goods.

**Collaboration and team dynamics**: Feedback from the experts who led the roundtable discussion highlighted positive group dynamics, with only a few minor conflicts observed. Participants valued the interactive and hands-on aspects of the Ideathon, expressing satisfaction with the collaborative process. Attendees were enthusiastic and actively engaged.

**Community engagement and post-event impact**: The municipality of Pörtschach found valuable insights among the proposed solutions, particularly those focused on integrated transportation of both passengers and goods, which align well with regional development initiatives. Efforts are underway to continuously inform Ideathon participants about upcoming AV deployments, enabling them to understand their role in shaping these developments.

**Challenges**: Overall, the Ideathon proceeded smoothly without major challenges, aside from some minor format and presentation issues related to the use of slides and digital devices. Initially, the participants were encouraged to use traditional analogue methods (e.g., on paper). Looking back, it might have been advantageous to enable the use of PowerPoint alongside analogue presentation techniques for the ease of documenting and preserving results.

## 4   Hackathon (Thessaloniki, Greece)

A SHOW Hackathon took place in Thessaloniki from 21st to 23rd March 2022, organized by CERTH (Centre for Research and Technology Hellas) with the support of Y4PT (Youth for Public Transport). Participants could be individuals of any age and professional background, who would demonstrate their commitment, creativity, or enthusiasm for sustainability in their application. In total, 20 participants joined the Hackathon, including students, software developers and young researchers, working in teams. Experts from the SHOW consortium and the global Y4PT team were at hand throughout the three days to guide and mentor the teams.

The objective of the Hackathon was to develop added value services to meet end-users' (unfulfilled) needs related to the use of CCAM, focusing on three topics selected from a previous project Ideathon, namely:

1. Human assistance stand-by in case of problems. People may feel unsafe using an automated vehicle if there is no safety driver. Cameras on board could automatically detect a problem (using Artificial Intelligence) and notify the control centre which, if needed, can send someone to help.
2. Adapting capacity to increase in demand in a flexible way. Surplus demand (e.g., due to peak hour, large events, unforeseen circumstances) can be addressed by quickly deploying additional automated buses, to be summoned from a nearby bus depot.
3. Accessibility and assistance for persons with reduced mobility (PRM). How can we ensure that PRM can travel independently, how can we keep them informed

(e.g., through audio-visual messaging) about accessibility, and how can they notify any need for assistance?

The three finalist teams enthusiastically took on the task to design and develop novel solutions and applications for each of the challenges:

- *Team 1—Gustav*: tackled the issue of making driverless automated vehicles safer and more secure for passengers and people on the road.
- *Team 2—DeFORUS*: worked on a project for adapting/ optimising capacity to handle demand in a flexible way.
- *Team 3—AsistIO*: addressed the important issue of accessibility and assistance for persons with reduced mobility.

The evaluation focused on meeting expected outcomes and specifications for each challenge, considering criteria such as innovation and creativity, functionality, impact, feasibility of implementation, inclusivity, and presentation and communication skills.

Regarding the effectiveness of the method:

**Quality of insights**: The ideas cultivated during the Hackathon demonstrated a remarkable commitment to innovation, particularly in addressing unmet needs associated with the use of CCAM technologies, focusing on the three targeted topics. Participants showed a keen understanding of the end-users' challenges and strived to develop added-value services that directly catered to these needs. Through collaborative efforts and creative problem-solving, teams proposed solutions aimed at enhancing the relevant CCAM experience, whether by increasing flexibility in demand, optimizing safety protocols, or improving accessibility for diverse user groups. The depth of insight and ingenuity displayed throughout the event underscored the potential for transformative advancements in CCAM services, poised to enhance both user satisfaction and overall system efficiency.

**Collaboration and team dynamics**: The Hackathon participants provided positive feedback, expressing appreciation for the well-organized event and the opportunity to collaborate with diverse and skilled individuals. They commended the clear communication of challenges, the availability of experienced mentors and the seamless coordination that facilitated a smooth workflow. Many participants highlighted the valuable learning experiences gained during the Hackathon and praised the inclusive and supportive environment fostered by the organisers. Effective collaboration among team members from diverse backgrounds proved highly successful, resulting in innovative solutions that capitalized on each team member's strengths.

**Community engagement and post-event impact**: Due to their high quality, all three ideas/concepts have been endorsed and promoted through the SHOW project Marketplace (SHOW 2024). Moreover, *Gustav* has led to a series of follow-up implementations, under different configurations, in the Trikala and Madrid SHOW pilot sites, whilst future implementations are expected in the ULTIMO project.[2] *DeFORUS* has led to different AI services configurations, based on only historical or historical

---

[2] https://www.ultimo-he.eu/.

and real time/near real-time fed data. Specifically, on a historical data basis, similar services have been deployed in the Tampere and Frankfurt SHOW pilot sites, while on a historical and real time/near real-time data feeds basis, implementations have been/are being followed in the Trikala, Madrid, Turin and Carinthia pilot sites.

**Challenges**: However, the organizing team also identified challenges. While collaboration was a strength, there is a need to further enhance diversity and inclusivity in future Hackathons, ensuring equal participation and representation across different demographics. Participants expressed a desire for continued post-Hackathon support, including access to resources, networking opportunities and mentorship. Establishing a more robust feedback mechanism for both participants and mentors became a key area for improvement.

## 5  MAMCA Workshop (Tampere, Finland)

The objective of the MAMCA approach in SHOW was to evaluate the impacts of the different automation scenarios tested in the project from multiple stakeholder perspectives. We discuss below the application of the MAMCA methodology in the SHOW pilot site in Tampere, Finland conducted on March 17th 2023.

A total of 24 participants were invited to take part in the Tampere MAMCA workshop, each representing a stakeholder group which has an interest and stake in the topic. Table 2 below summarizes the stakeholder groups, scenarios and criteria used in the Tampere workshop.

While the MAMCA workshop is relatively short in comparison to other user engagement approaches (less than half a day), it does require significant preparation ahead of the day of the workshop. In the case of the SHOW project, the MAMCA workshops were conducted by one entity (VUB), and the organization was done in collaboration with the local pilot site.

Preparation began 3 months before the planned date, the first steps being to identify the relevant stakeholders and to find a suitable date for the workshop. The hosts were expected to provide a venue, laptops for each stakeholder group and catering, and

**Table 2**  Summary of Stakeholder Groups, Scenarios, and Criteria used in MAMCA workshop—Tampere site

| Stakeholder groups | Scenarios | Criteria |
|---|---|---|
| • Vehicle users<br>• Local authorities / regulators<br>• Mobility service providers / public transport operators<br>• Public Interest Groups and Associations<br>• Research and Development | • Scenario 0: Business as Usual<br>• Scenario 1: Driverless shuttle for first/ last mile<br>• Scenario 2: Door-to-door delivery of persons and goods<br>• Scenario 3: Mass transit AV services<br>• Scenario 4: Shared robotaxis | • Road safety<br>• Traffic efficiency<br>• Energy efficiency<br>• Environmental impact<br>• Socio-economic impact<br>• Employment<br>• Social equity<br>• User acceptance |

to support the participants with travel arrangements if needed, while the organizing team from VUB prepared the workshop content (supporting documents, preparation of the MAMCA online tool etc.). The pilot site also provided translation of supporting documents and moderators who could support the groups in the breakout sessions.

The workshop is qualitative in nature, as it is focused on stakeholders' opinions and evaluation of the outlined scenarios. In the case of SHOW these scenarios were:

1. Business as usual
2. Driverless shuttle for first / last mile
3. Door-to-door delivery of persons and goods
4. Mass transit AV services
5. Shared robotaxis.

The stakeholders' opinions are reflected with scores entered by the stakeholders themselves on the dedicated MAMCA online tool, using the criteria mentioned in Table 2. This includes, per stakeholder group, criteria weights on a scale of 0–100, later normalized, as well as evaluation scores for each scenario per criterion from 1 to 10. A final aggregated score is then calculated for each stakeholder group and scenario using a weighted sum approach. This is applied for each scenario for each stakeholder.

The results for the Tampere workshop are visualized as depicted in Fig. 1.

Beyond the scores, discussions both in the stakeholder breakout groups (through moderator notes) and in the plenary discussion are recorded. Such data is not as structured as the scores shown above, but it provides additional context to interpret the results, as well as the process of assigning the weights and scores.

The overall result of the MAMCA workshop in Tampere shows a generally positive view of the shared autonomous mobility scenarios that were presented (see Fig. 1). Two scenarios emerged as the most positively rated: mobility service providers / public transport operators, public interest groups and researchers preferred *mass transit AV services*, while users and local authorities favoured *first / last mile*

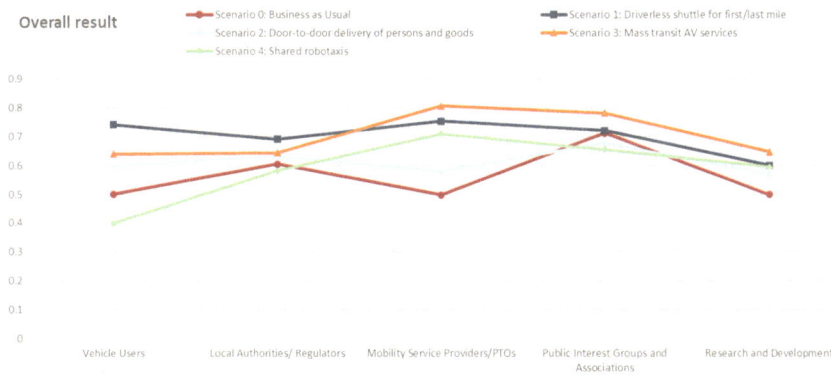

**Fig. 1** Overall MAMCA result—Tampere site

*driverless shuttles*. The *shared robotaxi* scenario, interestingly, was scored relatively low by most stakeholders and was not the preferred scenario for any of them. It was even rated lower than the *business-as-usual* scenario by users, local authorities, and public interest groups. The users' group—which included wheelchair users as well as people with a vision or hearing impairment—highlighted two reasons for this. Firstly, the usage of robotaxis would lead to more vehicles on the road with lower capacity compared to public transport, resulting in more traffic congestion. Additionally, the small size in comparison to buses and the absence of people in the vehicles pose significant challenges for users with a mobility, hearing or vision impairment, as they could need assistance. Because the users' group considered access to the service, comfort inside the vehicle, and clarity of information as critical factors in their weighting of the criteria, the robotaxi scenario had a low aggregate score.

Regarding the effectiveness of the MAMCA workshop method:

**Quality of insights**: The MAMCA workshop was used as a stakeholder analysis approach in the later phase of the innovation process. Thus, its aim was not to produce new ideas, but rather project into the future and reach consensus within each stakeholder group on how they would expect future scenarios of shared automation to perform. Its outcomes were then not completely unexpected and were rather a validation of insights. One of the most important insights is that there are trade-offs to be made when making decisions about automation scenarios, as often no scenario will perform well in all criteria, and they highly depend on the objectives of the stakeholders.

**Collaboration and team dynamics**: Overall, the feedback from participants of the Tampere MAMCA workshop was positive, with many pointing out the value of involving all stakeholders together in an open setting, allowing to share and discuss different perspectives in a structured manner. Specifically, users with different mobility needs (people with a vision or hearing impairment, wheelchair users) appreciated that they were invited to actively take part in such a discussion and share their concerns and opinions.

**Community engagement and post-event impact**: The participants identified several positive aspects of the workshop. First the structure of the workshop, starting with the in-group discussions and finishing with the plenary discussion with the immediate visual representation of the results, was appreciated as it provided a clear framework to guide the discussion. The translated handouts with simplified explanations of the scenarios and criteria were also helpful to the participants. Additionally, we found that engaging with the hosts early in the content preparation and briefing the moderators ahead of the workshop was useful as they could provide further support in the group discussions, notably in the local language. The feedback from users with different mobility needs encouraged the pilot site leaders to have more direct discussions with them in the future. The results of the workshop also validated the city authorities' decision to focus on autonomous public transport applications and emphasized the need for consultation with other stakeholders and understanding how they perceived the different autonomous mobility alternatives.

**Challenges**: A commonly reported challenge was the difficulty of estimating some impacts in the future scenarios. A recurring example was employment, as stakeholders often struggled to visualize a single trend for employment in future scenarios of automation. Nonetheless, this provided ground for discussion within the groups and in plenary. Another challenge to consider during the stakeholder identification phase is to ensure that a single stakeholder group remains somewhat homogeneous. Indeed, if one stakeholder group is composed of multiple organizations, it is important to make sure that there is reasonable convergence of objectives to avoid deadlocks.

# 6   Discussion and Recommendations

Whereas the importance of stakeholder engagement is increasingly recognized— an "integrated and participatory approach" is for example a key element in the development of Sustainable Urban Mobility Plans (SUMPs) (Recommendations on National Support Programmes for Sustainable Urban Mobility Planning 2023)— real co-creation in transport planning seems to be still a rather new phenomenon, as reflected by a low volume of academic articles to be found on the subject (Pappers et al. 2020). Nevertheless, European projects / initiatives can be seen as the ideal platform to bring stakeholders together with researchers and citizens for developing and deploying inclusive and meaningful mobility products and services. The topic of citizen and stakeholder engagement has already been addressed in several European-funded and CCAM-related projects, in many cases with a direct link to user acceptance, e.g., PAsCAL,[3] SUaaVE[4] and Drive2theFuture.[5]

Within SHOW, a large number of stakeholder and citizen engagement activities have been conducted, including creative formats that bring participants together in person for a defined period of time. Diverging from more traditional methods like surveys, the engagement formats as presented in this paper—Ideathon, Hackathon, MAMCA workshop—embrace a collaborative and interactive approach, bring together participants from diverse backgrounds and stimulate them to think outside of the box by adopting creative techniques, thus encouraging the generation of innovative ideas and solutions within a short timeframe.

In particular, our review showed that:

The **quality of ideas or insights** was the highest for the Hackathon. Whereas the focus of an Ideathon is on the development of ideas or concepts rather than the creation of finished products or solutions, a Hackathon aims at producing tangible outcomes, such as prototypes, proofs of concept or functional solutions. This is corroborated by the swift implementation of the SHOW Hackathon's outcomes in some of the SHOW pilot sites.

On the other hand, the lowest level of originality was within the MAMCA workshop, as the workshop was designed to support decision making and consensus

---

[3] https://www.pascal-project.eu/.

[4] https://www.suaave.eu/.

[5] https://www.drive2thefuture.eu/.

building. In the end, the method—embedded towards the end of the innovation process—was about highlighting the different perspectives of the stakeholders and providing a multi-actor evaluation of automation scenarios.

The combination of activities works well too: An Ideathon is a good way of identifying first ideas on possible solutions to a challenge, while a Hackathon can further develop these ideas towards an actual 'product' ready to be deployed.

For all three methods, gathering participants in person has shown to be effective for **collaboration and team dynamics**. Overall, involving all stakeholders including end-users can positively contribute to a higher acceptance and a more positive evaluation of new CCAM solutions. Both the participants and the organisers of the SHOW Ideathon, Hackathon and MAMCA workshop gave positive feedback. Participants praised the opportunity to work on solutions for the mobility of the future as non-experts. In the MAMCA workshop, for example, users with disabilities appreciated being invited to actively take part in such a discussion and share their concerns and opinions.

Regarding **post-event impacts**, both the participants and organisers found the results useful for future projects and operations due to their high quality. As such, the outcomes of all three methods have been and will be utilised in future activities within SHOW, while concepts from the Hackathon will also be implemented in the ULTIMO project.[6] While stakeholders want to obtain useful results (innovative ideas, feasible solutions, usable input for future policy decisions, etc.), participants want to know how their input will be considered and how their ideas or solutions will be implemented.

As for **challenges**, organising successful citizen or stakeholder engagement events unavoidably takes time and requires effort and resources. A good preparation is key, both in terms of practical aspects (e.g., catering and supporting materials), support during the event, and content-wise (defining objectives).

No matter the stage within the innovation process (idea generation, development or validation), co-creative methodologies require a professional approach. Moreover, it is crucial to pay sufficient attention to recruiting a diverse range of participants, which in turn requires providing them with enough incentives to be motivated to participate. In terms of evaluation, we recommend to think beforehand about evaluation criteria in line with the goals to be achieved, making sure that the outcomes and results are judged fairly and transparently.

We encourage new research to further explore the application and possible combination of citizen and stakeholder engagement methodologies to enhance our understanding of the needs of people and practitioners at the local level. This will ensure that CCAM solutions are developed and deployed accordingly.

---

[6] https://ultimo-he.eu/.

# 7 Conclusion

In conclusion, our study emphasizes stakeholder engagement's rising significance in planning CCAM, with participatory methods promising innovation and inclusive decision-making. We showed that diverse activities like Ideathons, Hackathons and MAMCA workshops efficiently generated ideas and helped validate local solutions with novel insights and perspectives, yet resource allocation and participant diversity posed challenges. Future research should focus on standardizing engagement methodologies for better cross-context comparison, aiding evidence-based decision-making and effective approaches in developing and deploying CCAM.

**Acknowledgements** The research is part of the SHOW project that has received funding from European Union's Horizon 2020 research and innovation programme under grant agreement no. 875530.

# References

Arnstein S (1969) A ladder of citizen participation. J Am Plann Assoc 35(4):216–224

Chng S, Kong P, Lim PY, Cornet H, Cheah L (2021) Engaging citizens in driverless mobility: Insights from a global dialogue for research, design and policy. Transp Res Interdisc Perspect 11:100443 [Online]. Available: https://doi.org/10.1016/j.trip.2021.100443

Cornet Y, Barradale MJ, Gudmundsson H, Barfod MB (2018) Engaging multiple actors in large-scale transport infrastructure project appraisal: an application of MAMCA to the case of HS2 High-Speed Rail. J Adv Transp 1–22. https://doi.org/10.1155/2018/9267306

Feys M, Rombaut E, Macharis C, Vanhaverbeke L (2020) Understanding stakeholders' evaluation of autonomous vehicle services complementing public transport in an urban context. In: 2020 forum on integrated and sustainable transportation systems (FISTS). IEEE, pp 341–346. https://doi.org/10.1109/FISTS46898.2020.9264856

Grandsart D, Cornet H, Loukea M, Coeugnet-Chevrier S, Metayer N, Anund A, Sjörs Dahlman A (2023) Citizen and stakeholder engagement in the development and deployment of automated mobility services, as exemplified in the SHOW project. In: Nathanail EG, Gavanas N, Adamos G (eds) Smart energy for smart transport. Springer, Cham, pp 465–476 [Online]. Available: https://doi.org/10.1007/978-3-031-23721-8_39

Hackathon.com, What is a Hackathon? [Online]. Available: https://tips.hackathon.com/article/what-is-a-hackathon. Accessed 22 Dec 2021

Huang H, Lebeau P, Macharis C (2020) The Multi-Actor Multi-Criteria Analysis (MAMCA): new software and new visualizations. In: Moreno-Jiménez JM, Linden I, dargam f, jayawickrama u (eds) decision support systems x: cognitive decision support systems and technologies. Springer International Publishing, pp 43–56. https://doi.org/10.1007/978-3-030-46224-6_4

Jones GM, Semel B, Le A (2015) There's no rules. It's hackathon: negotiating commitment in a context of volatile sociality. J Linguist Anthropol 25(3):322–345

Lebeau P, Macharis C, Van Mierlo J, Janjevic M (2018) Improving policy support in city logistics: the contributions of a multi-actor multi-criteria analysis. Case Stud Transp Policy 6:554–563. https://doi.org/10.1016/j.cstp.2018.07.003

Leven M, Kersten W (2021) Integration of stakeholders in the innovation process of transportation networks. Hamburg Int Conf Logist (HICL) 2021:749–766

Macharis C, de Witte A, Ampe J (2009) The multi-actor, multi-criteria analysis methodology (MAMCA) for the evaluation of transport projects: theory and practice. ATR 43:183–202. https://doi.org/10.1002/atr.5670430206

Macharis C (2005) The importance of stakeholder analysis in freight transport. Euro Transp\Trasporti Europei 114–126

Official Journal of the European Union (2023) Recommendations on national support programmes for sustainable urban mobility planning, Commission Recommendation (EU) 2023/550, notified under document C(2023) 1524

Pappers J, Keserü I, Macharis C (2020) Co-creation or public participation 2.0? an assessment of co-creation in transport and mobility research. In: Müller B, Meyer G (eds) Towards user-centric transport in Europe 2. Springer, Cham, pp 3–15. https://doi.org/10.1007/978-3-030-38028-1_1

Rowe G, Frewer LJ (2005) A typology of public engagement mechanisms. Sci Technol Human Values 30(2):251–290

SHOW (2020) Deliverable 1.1 Ecosystem actors needs, wants and priorities and user experience exploration tools, European Commission Horizon 2020 GA 875530

SHOW Marketplace [Online]. Available: https://show-project.eu/ccam-marketplace. Accessed 29 March 2024

Sakiyama M, Fujii N, Kokuryo D, Kaihara T (2020) Visualization of group discussion using correspondence analysis and LDA in Ideathon. Kobe University, 1–1, Rokkodai-cho, Nada ward, Kobe City, Hyogo prefecture, Japan, Elsevier B.V

Sun H, Zhang Y, Wang Y, Li L, Sheng Y (2015) A social stakeholder support assessment of low-carbon transport policy based on multi-actor multi-criteria analysis: the case of Tianjin. Transp Policy 41:103–116. https://doi.org/10.1016/j.tranpol.2015.01.006

Verworn B, Herstatt C (2002) The innovation process: an introduction to process models, Working paper, No. 12

# Correlation of Shared Automated Vehicles Real Traffic Performance and Passengers' Acceptance Data

**Alexandros Papadopoulos, Georgios Spanos, Jordi Pont, Antonios Lalas, Konstantinos Votis, Maria Gkemou, Anna Anund, Karl Lambauer, Lucia Isasi De La Iglesia, Dimitrios Tzovaras, and Evangelos Bekiaris**

**Abstract** The perception of comfort and safety among passengers of Autonomous Vehicles (AVs) is crucial and significantly influences their adoption in current Public Transport systems. It is essential to align the objective perception with an analysis of vehicle performance data to identify vulnerabilities and factors affecting passenger comfort and safety. This paper presents the first comprehensive correlation between objective and subjective data from autonomous fleets in three well-established pilot

A. Papadopoulos (✉) · G. Spanos · A. Lalas · K. Votis · D. Tzovaras · E. Bekiaris
Centre for Research & Technology Hellas (CERTH), Information Technologies Institute (ITI), 6th Km Charilaou-Thermis, 57001 Thessaloniki, Greece
e-mail: alexpap@iti.gr

G. Spanos
e-mail: gspanos@iti.gr

A. Lalas
e-mail: lalas@iti.gr

K. Votis
e-mail: kvotis@iti.gr

D. Tzovaras
e-mail: tzovaras@iti.gr

E. Bekiaris
e-mail: abek@certh.gr

J. Pont
Electronics, IDIADA Automotive SA, Ronda de L'Albornar S/N, Santa Oliva, 43710 Tarragona, Spain
e-mail: jordi.pont@idiada.com

M. Gkemou
Centre for Research & Technology Hellas (CERTH), Hellenic Institute of Transport (HIT), 34 Ethnarchou Makariou St., 16341 Athens, Greece
e-mail: mgemou@certh.gr

A. Anund
Swedish National Road and Transport Research Institute, Olaus Magnus Väg 35, 581 95 Linköping, Sweden
e-mail: anna.anund@vti.se

© The Author(s) 2025
H. Cornet and M. Gkemou (eds.), *Shared Mobility Revolution*, Lecture Notes in Mobility, https://doi.org/10.1007/978-3-031-71793-2_11

locations (Graz, Madrid, Linköping), each using different technologies and experiencing varying environmental conditions. Our analysis (i) revealed significant differences between the three pilot sites in terms of perceived safety and comfort (both perceived and actual) and (ii) confirmed a strong correlation between safety and comfort levels and the vehicles' behaviour in terms of speed and acceleration, particularly noting the impact of hard braking events as those were defined by the SHOW consortium.

**Keywords** Cooperative Connected Automated Mobility (CCAM) · Real traffic · Performance data · Safety perception · Comfort perception · Correlation analysis

# 1 Introduction

The automotive industry stands on the cusp of a transformative era driven by advancements in technology, automation, and data-driven decision-making. In this dynamic landscape, ensuring the safety and comfort of drivers and passengers remains paramount.

While safety comes undoubtedly first, which turns the goal of providing safe transport services of paramount importance, it is quite often that the perception of safety on behalf of the (whichever) technological system users is not necessarily aligning with that (Moody et al. 2020). Users' perception of safety is defined by their awareness of the risks taken and calculated. Passengers feel insecure when the risks can either not be predicted or exceed the subjective thresholds pre-defined by each individual (Friman et al. 2020). Even in the case that perceived and actual safety are aligning, the perception of comfort with regard to experienced services often differentiates overall. Comfort in public transport is largely governed by globally recognized standards, such as the EN 13,816 (European Committee for Standardisation 2002) standard endorsed by the European Union. This standard evaluates passenger satisfaction across a range of factors including convenience, accessibility, information availability, travel time, customer service, comfort, security, and environmental impact (İmre and Çelebi 2017).

In the transport field, the entire value chain, consisting of manufacturers, service providers and researchers, works from different edges on the same objectives

K. Lambauer
Embedded Systems Group (ESG), VIRTUAL VEHICLE Research GmbH (VIF), Inffeldgasse 21/A, 8010 Graz, Austria
e-mail: Karl.Lambauer@v2c2.at

L. I. De La Iglesia
Fundación Tecnalia Research & Innovation, Industry and Mobility, Astondo Bidea, Edificio 700, 48160 Derio (Vizcaya), Spain
e-mail: lucia.isasi@tecnalia.com

pursuing ultimately to increase efficiency and acceptance of the technological solutions. These stakeholders are very often exploring both operational performance of vehicles and user's perception of their operation. This approach gets more valuable when is applied in novel mobility paradigms, such as the shared automated mobility one.

In the SHOW project (2024), the evaluation and impact assessment protocol established was both subjective and performance data driven (SHOW 2020) in the sense that it tackled the safety and efficiency assessment of the Connected and Cooperative Automated Mobility (CCAM) services provided across its pilot sites, exploring at the same time how those services have been subjectively assessed by the passengers and the full spectrum of stakeholders involved (i.e., safety drivers, remote operators, transport operators, etc.). While the full approach applied in SHOW entails a long list of measuring tools and evaluation means and methodologies to tackle all different aspects and layers of anticipated and not anticipated impacts in mobility, traffic safety, traffic efficiency, user experience and acceptance and in the energy consumption and the environment, the current manuscript deals specifically with the topic of correlating the performance of automated vehicles and their passengers' perception of their performance against two specific aspects, namely perceived safety and comfort along specific time slots of ridership selected for three pilot sites of SHOW, namely Graz in Austria (Graz site 2024), Linköping in Sweden (Linköping site 2024) and EMT bus depot in Carabanchel in Madrid, Spain (Madrid site 2024).

The objective of the analysis performed and presented herein—and which has been applied for most of the pilot sites in SHOW project—is to gain deeper insights into the correlation of passengers' perception of safety and their experienced comfort with the quantified performance of the Connected Automated Vehicles (CAVs) deployed in specific shared mobility services, as they are presented below. By bridging the gap between actual and perceived performance, stakeholders (Original Equipment Manufacturers, Tier 1 suppliers and transport operators) can make informed decisions to enhance vehicle design, functionality, and, finally, user experience and acceptance.

The remainder of this paper is structured as follows: Sect. 2 reviews related studies and Sect. 3 describes the methodology employed. In Sect. 4, we present the SHOW ecosystem along with the properties and technologies used at the pilot sites. Section 5 details the analysis of performance and subjective data, and also discusses the correlation with hypotheses and general findings. Finally, Sect. 6 presents the conclusions of the paper.

## 2 Relevant Studies

Perceptions of safety and comfort are subjective metrics that are influenced by vehicle performance. The connection between these factors is crucial in both conventional and automated vehicles. Especially in automated vehicles, achieving high safety and comfort levels may be even more challenging due to the continuous evolution of the technology.

The study cited as Asua et al. (2022) explores how advanced driving assistance systems impact passenger comfort and susceptibility to motion sickness. It evaluates passenger acceleration data across various driving styles, vehicle types, and road conditions, revealing that the type of road significantly impacts comfort. Additionally, the research in Eboli et al. (2016) introduces a method to classify driving behaviours as safe or unsafe using kinematic parameters such as speed and acceleration. This study uses a smartphone app to gather geo-referenced data from vehicles on a rural two-lane road.

Kumar et al. (2018) investigate the vertical dynamics of a passenger car to assess ride comfort using bond graph modelling. The research takes into account various factors such as the suspension system, seat, sprung and not-sprung masses, and the vertical motion of passengers. It evaluates comfort by comparing weighted root mean square acceleration at the passenger seat against ISO-2631 standards. Lastly, the authors in Atarod (2021) analyze the occupant dynamics in side impact tests across various vehicle models and years. It is therein assessed the safety of occupants by evaluating how dynamic responses like head acceleration, shoulder lateral forces, and spine accelerations correlate with injury assessment reference values. The study provides insights into the changes in vehicle safety features over time and their effectiveness in protecting occupants during side impacts.

To our knowledge, there is no research that correlates subjective and vehicle performance data in the field of automated vehicles. Analyzing performance data alongside questionnaire surveys results could help identify the acceleration and speed profiles that ensure higher levels of comfort and safety. Additionally, this analysis could highlight the actual vulnerabilities of the autonomous vehicle system, distinguishing them from possible trust issues associated with this innovative technology.

## 3   Methodology

The correlation of vehicle performance and subjective data in the context of the SHOW project involves examining the relationship between quantitative performance metrics and qualitative user experiences. This analysis seeks to understand how objective performance indicators, such as speed profiles, acceleration profiles, and the frequency of hard brakes, align with subjective assessments of user comfort and safety perception during test runs in SHOW pilot sites. While objective measures provide valuable insights into the technical capabilities of CAVs, subjective measures are crucial for understanding the overall user experience and acceptance of these vehicles. Understanding the correlation between performance and subjective data is essential for optimizing the performance of automated vehicles (AVs). By analyzing how objective performance metrics relate to subjective user experiences, researchers and developers can refine AV systems to maximize user satisfaction and acceptance. This can lead to the development of safer, more comfortable, and more widely accepted AVs that seamlessly integrate into daily lives.

Our methodological approach encompasses data collection, preprocessing, hypothesis formulation, statistical analysis, and the interpretation of results as described below:

1. **Data collection**

    (a) **CAV Performance data**. The first data source derives from the sensors mounted on the AVs. The analysis demands only two main attributes that are broadly available from all the manufacturers: timestamp and speed value. These minimum requirements ensure the robustness of the analysis and its usage in any AV.

    (b) **Acceptance survey data**. The other source of data is the acceptance survey data collected through a dedicated platform in the project (Netigate 2024) and via common acceptance surveys designed to be conducted in all project pilot sites. These surveys were administered to passengers and other stake-holders participating in the pilots and, in the case of passengers, it sought to capture their subjective experiences during test rides. Among all aspects addressed therein, the focal aspects for the current work are on two key aspects sought in the context of those surveys, namely comfort level and safety perception. Those formulated the subjective basis for this analysis.

2. **Data pre-processing**. The performance data was then processed to extract relevant metrics, namely the acceleration values, the speed/acceleration profiles and the identification of instances of hard braking events. As concerns the acceptance survey data, it was processed and organized to ensure data consistency. Any missing or erroneous responses were manipulated respectively to maintain data quality.

3. **Hypotheses testing**. Building on previous research and expertise from projects like SUAAVE[1] and AUTOPILOT,[2] initial hypotheses were formulated regarding the expected correlations between subjective passenger perception and vehicle performance data. These hypotheses include:

    (a) A smoother speed profile correlates with higher passenger comfort levels.

    (b) A smoother speed profile correlates with higher safety perception levels.

    (c) A lower number of hard braking events correlates with higher passenger comfort levels.

    (d) A lower number of hard braking events correlates with higher safety perception levels.

    (e) There is no statistical difference regarding the actual comfort among the pilot sites.

    (f) There is no statistical difference regarding the actual safety among the pilot sites.

    (g) There is no statistical difference regarding the comfort perception among the pilot sites.

---

[1] https://www.suaave.eu/news/.

[2] https://autopilot-project.eu/.

**Table 1** Categorization of level of comfort with the acceleration values

| Acceleration range | Description |
|---|---|
| >3 m/s$^2$ | Hard braking events |
| 2–3 m/s$^2$ | Extremely uncomfortable |
| 1.25–2.5 m/s$^2$ | Very uncomfortable |
| 0.8–1.6 m/s$^2$ | Uncomfortable |
| 0.5–1 m/s$^2$ | Fairly uncomfortable |
| 0.315–0.63 m/s$^2$ | Little uncomfortable |
| <0.315 m/s$^2$ | Comfortable |

(h)  There is no statistical difference regarding the safety perception among the pilot sites.

4. **Thresholds for comfort**. To assess passenger comfort levels objectively, established thresholds were referenced. In this analysis, we employ the following methodology to categorize acceleration values into corresponding levels of passenger comfort. The methodology has been introduced by Asua et al. in their recent work in 2022 (Asua et al. 2022). In their paper, the authors have made an endeavour to establish comfort levels for AVs. Through a comprehensive statistical analysis, they have defined the acceleration thresholds, as presented in Table 1. Based on the analysis implemented in terms of SHOW project, we have added the definition of the hard-braking events.

5. **Thresholds for safety levels.** The safety levels of the autonomous fleets are assessed using the methodology outlined in Eboli et al. (2016). This research presents a method to distinguish between safe and unsafe situations in a conventional vehicle. The approach involves categorizing these scenarios based on acceleration thresholds, which are derived from the vehicle's speed using the formula:

$$|a| = g * \left[ 0.198 * \left( \frac{V}{100} \right)^2 - 0.592 \left( \frac{V}{100} \right) + 0.569 \right]$$

The definition of the acceleration thresholds allows for the creation of two distinct curves. By plotting actual acceleration values on the same graph, one can visualize the perception of safety. This perception is quantified by the proportion of actual values that fall between the two threshold values.

6. **Statistical analysis**. For the aforementioned hypotheses testing, various appropriate statistical tests are performed depending on the data and the kind of hypotheses. More specifically, for the first four hypotheses the parametric Pearson correlation (Armstrong 2019) is conducted in order to correlate actual performance values such as speed (as reflected by coefficient of variation Jalilibal et al. 2021) and hard braking events with subjective passenger perception in terms of safety and comfort. Similarly, for the last four hypotheses, regarding the

actual and subjective comfort and safety levels among sites, the non-parametric Kruskal–Wallis (McKight and Najab 2010) and Mann–Whitney U (MacFarland et al. 2016) tests are employed. These statistical tests assess differences among three or more groups in the case of Kruskal–Wallis and between two groups for Mann–Whitney U.

7. **Ordinal Values Transformation**. In order to test hypothesis "e", the following approach to transform the ordinal variables of the dataset to numerical (Toloudis et al. 2016) was adopted. More specifically, each acceleration value corresponding to a comfort qualitative rating from "Not Uncomfortable" to "Hard Braking Events" has been associated with a respective numerical value. Hence, the "Comfortable" rating corresponds to 1, the "Hard Braking Events" corresponds to 7 respectively, while all the intermediate ones align to the respective numerical values (2–6).

# 4   SHOW Operating Sites in a Nutshell

There are three pilot case studies selected out of SHOW project to present the methodology applied as well as the derived outcomes, namely the robo-taxi pilot service deployed in open traffic in Graz, Austria, the shuttle service deployed in open traffic in Linköping, Sweden and the bus pilot service deployed in the confined environment in the EMT bus depot in Carabanchel, Madrid, Spain. Those have been selected as they have deployed different types of shared AV services, deploying also different types of AVs. Their pilot automated services are shortly described below.

**Graz pilot site, Austria**

The pilot at the Graz site was carried out with the focus on connecting peri-urban regions to intermodal mobility hubs in mixed traffic. The pilot was located on the southern outskirts of the city, about 4 km from the city centre. In this area, there are a shopping centre and the public transport (PT) hub "Puntigam", around 1 km apart and connected via the VIF-robotaxis. The PT hub "Puntigam" is served by six (6) public bus lines and 1 tram line, and is also connected to the nearby urban railway stations. In this urban scenario, the robotaxis stopped at the bus terminal, picked up people upon request, drove through the public stops, where there were many pedestrians, and then arrived the shopping centre. With the help from the traffic infrastructure (e.g., guiding through traffic lights), vehicles could perform actions in an automated way (at slow speed). Two (2) robo-taxis were in use at this site, i.e. one Ford Fusion and one Kia e-Soul. The Kia Soul from AVL was updated to support the same capabilities (automated driving software) as the Ford Fusion. The deployment of the infrastructure components (awareAI camera, Cooperative—Intelligent Transport System (C-ITS) Road Side Unit) was also carried out, configured and tested (Yunex). Besides the 4G/5G network, Vehicle-to-Everything (V2X) communication based on ITS G5 was used, setup in close cooperation with KAPSCH TrafficCom and Yunex. The public pilot phase lasted twelve (12) months, from October 2022 to September

2023. Around 520 passengers were transported in total. The passenger rides were conducted during public weeks and other events open to all visitors.

### Linköping pilot site, Sweden

The route was 4.2 km long and had thirteen (13) stops. The site was both at Linköping University and in the newly built residential area Vallastaden, built with sustainable smart city in mind and with relatively few parking spaces and optimized for walking and cycling. The site included both mixed traffic with separate lanes for Vulnerable Road Users (VRUs) and shared spaces with VRUs (Anund et al. 2022). The speed limit on the streets was between 30 and 40 km/h, but the speed of the shuttle was always less than 20 km/h. A multi brand approach was used with one Navya DL4 shuttle L4 and two EasyMile EZ10 Gen2 shuttles. The operations were ongoing for twenty-three (23) months from February 2022 to December 2023. The shuttle service was carried out from 8 am to 5.30 pm weekdays and 11 am–4.30 pm weekends. Users were able to easily see where the shuttles were with help of a digital map with real time data visualizing both the shuttles and PT, with the intention to support planning for a seamless journey where the shuttles offer a "first mile—last mile" service to existing public transport. In addition, customers could request a ride by sending information through the app or by pressing a button at the bus stop. In Linköping, the geographical context was considered important in terms of evaluating how the mobility service and its technology fit into a real-life context and, especially, with regard to interaction with other road participants.

### Carabanchel EMT bus depot, Madrid, Spain

This pilot site was set in Carabanchel bus depot, a semi-controlled area with interaction with other non-automated buses and vehicles, as well as daily operations at the depot (manoeuvring, moving goods, people, etc.). Hence, the traffic environment at this site was about equivalent to an urban one, with interaction with cars, buses, trucks and VRUs (pedestrians, such as employees and visitors). The final public phase operated for ten (10) months, from November 2022 to October 2023. The route included five (5) stops and a round trip of 800 m long connecting different facilities within the bus depot. The service successfully run in weekdays until the afternoon, making use of one Gulliver minibus and onc 12 m i2eBus. Wednesdays were a particular day at the depot as it is considered students' day visit. Taking advantage of this (almost) weekly event, the young generation dropping in on the depot had the opportunity to experience the 5-stops service project, while visiting the different depot areas (training, boxes, main building, etc.). Passengers stayed at the dedicated bus stops, waiting for the bus to arrive, and participated at the service. The speed limit within the bus depot was 10 km/h, dropping to 5 km/h in the bus cleaning area, and 20 km/h on the bus testing circuit. During the pilots, the CAVs reached a maximum speed of 20 km/h. The most important goal of the 5-stops service at the depot was to improve high automation driving experience both for the users (passengers) and the safety drivers, encompassed in the public transport sector.

In all three sites, the participating vehicle fleets shared occasionally vehicle performance data with the SHOW Data Management Portal (entity responsible for the performance data gathering and processing in the project), either in real time or off-line, upon specifically defined metrics, the ones utilized in this study included among other.

# 5 Analysis

The performance data was acquired through the sensors mounted on the vehicles. Among other metrics, all sensors are capable of providing real-time speed measurements in meters per second at predefined intervals. From this information, both the vehicle's speed and the corresponding timestamp were extracted. By calculating the time difference between two consecutive speed values, the respective acceleration values have been determined adopting the same simple approach in the related study of Papadopoulos et al. (2024). This methodology has also been employed in vehicles where acceleration data is available through a dedicated data capture mechanism, resulting in consistent values. The calculated acceleration values corresponding to the days on which the questionnaires were collected have been categorized according to the groups described in Sect. 3. Moreover, with respect to the subjective data, in terms of perceived comfort and safety, as reflected by the answers of the passengers, it is worth mentioning that there is a Linkert scale from 1 to 9, where the "9" represents the highest positive level either in comfort or in safety and the "1" the lowest one.

## 5.1 Case Study 1: Graz

The first case study focuses on the Graz pilot site, where subjective and performance data from June 2023 were analysed. We retained only the performance data corresponding to the dates on which the subjective data was collected. The relevant data for the Graz site spans **four days**. The total number of speed measurements, each with an accompanying timestamp, is **17,932**. The average speed recorded is **16.81** kms per hour (km/h), with a variance of **9.75** $km^2/h^2$, coefficient of variation (CV) equals **0.19** and max speed observed is **56.05** km/h. An instance from 14–06–2023 related to speed and acceleration values can be observed in Fig. 1.

Beginning with the analysis of the actual comfort levels, the respective percentage values per category, as depicted in Fig. 2, are **"Comfortable" 83.71%**, **"Little Uncomfortable" 10.33%**, **"Fairly Uncomfortable" 3.94%**, **"Uncomfortable" 1.6%**, **"Very Uncomfortable" 0.4%**, **"Extremely Uncomfortable" 0.02%** and **"Hard braking events" 0%**. As concerns the alignment of each level with the actual comfort rating, the Graz site has a median value equal to **1**, mean value equal to **1.24** and, lastly, the variance is **0.41**.

Regarding actual safety, Fig. 3 illustrates the acceptable acceleration thresholds alongside with the actual acceleration values, considering the speed values. Notably,

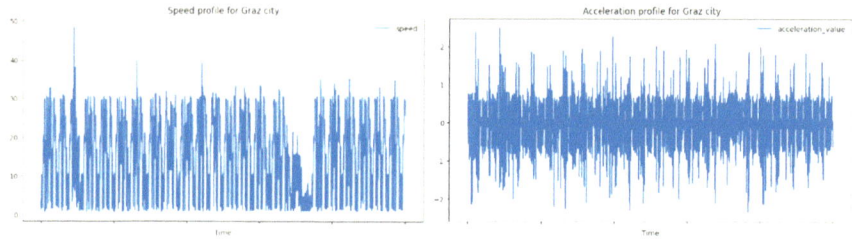

**Fig. 1** Speed (left) and acceleration (right) instance from Graz. *Source* Authors' own pictures

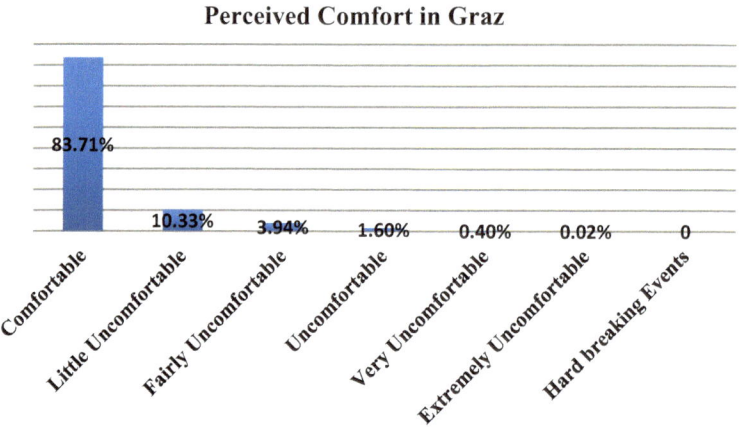

**Fig. 2** Perceived comfort in Graz. *Source* Authors' own pictures

no values fall outside the computed threshold areas with respect to the safety level. Additionally, the majority of the acceleration values cluster around the centre, staying well clear of the extremely acceptable values.

Continuing with the subjective ratings from passengers (Fig. 4), a significant **82.2%** of them reported a high level of comfort, rated at level **7 or above**, having a mean value of **7.7**. This is further supported by the fact that **97.98%** of the performance data falls within the zone up to '**Fairly Uncomfortable**'. However, the perception of safety is somewhat limited, with a lower mean value compared to comfort (7 and 7.7 respectively) and only about **60%** feeling secure (rate equals or greater to 7)—a finding that the performance data, which shows all acceleration values well within acceptable thresholds, does not explain. Additionally, there were **no instances of hard braking** that might reduce perceptions of safety. The only plausible explanation for this discrepancy is the combined total of **2%** for the '**Uncomfortable**', '**Very Uncomfortable**' and '**Extremely Uncomfortable**' categories, which is relatively high.

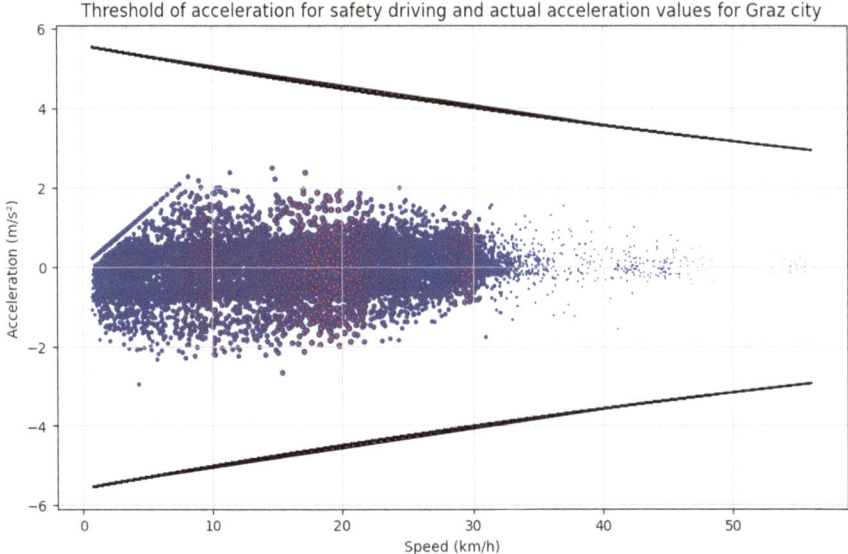

**Fig. 3** Safety acceleration profile for Graz. *Source* Authors' own pictures

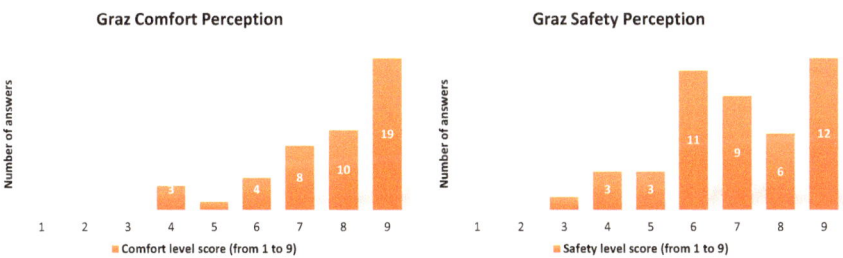

**Fig. 4** Ratings from passengers in Graz, related to comfort (left) and safety (right) perception. *Source* Authors' own pictures

## 5.2 Case Study 2: Linköping

The second case study refers to the Linköping pilot site. The analysed subjective and performance data were captured in September 2022. Similarly to the Graz case, the performance, and subjective data correspond to **four days** of operation. The amount of the speed and timestamp data are **74,204**. The mean value of speed is **7.51** km/h, the speed variance **3.26** km$^2$/h$^2$, the speed CV **0.24**, and the max speed **12.89** km/h. Some speed and acceleration instances from 10–09–2022 can be shown in Fig. 5.

Regarding the analysis of the actual comfort levels, the respective percentage values per category are **"Comfortable" 95.07%, "Little Uncomfortable" 4.47%,**

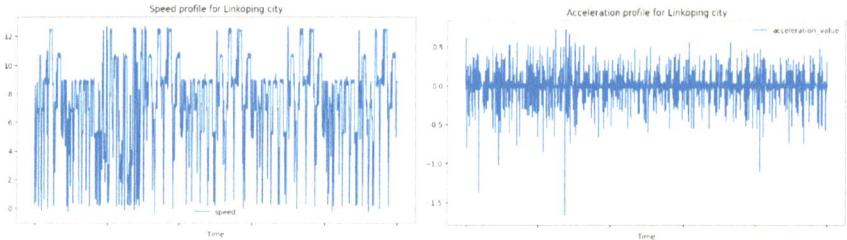

**Fig. 5** Speed (left) and acceleration (right) instances from Linköping. *Source* Authors' own pictures

**"Fairly Uncomfortable" 0.38%, "Uncomfortable" 0.05%, "Very Uncomfortable" 0.012%, "Extremely Uncomfortable" 0.013%** and **"Hard braking events" 0%** (Fig. 6). As concerns the alignment of each level with the comfort qualitative rating, the Linköping site has median value equal to **1**, mean value equal to **1.05** and, lastly, the variance is **0.06.**

In terms of the actual safety levels, Fig. 7 presents the acceptable acceleration thresholds in relation to the actual acceleration values. Similar to the previous findings from Graz, in Linköping no acceleration value exceeds the thresholds that objectively determine the safety limits. Compared to the Graz site, the values here are more centrally concentrated, maintaining a longer distance from the thresholds.

Finally, with respect to the analysis of the subjective surveys, the higher comfort passenger perception represents the **83.33%** (10 out of 12) of the sample, whereas the respective value for the safety perception is **100%** (Fig. 8). The very high comfort and safety levels are verified objectively by the no-existence of hard braking observations, the tiny total fraction of the "very" and "extremely uncomfortable" values that represent cumulatively only the **0.08%**, and the long distance of the data points from the acceptable acceleration thresholds. The smoother behavior of the Linköping

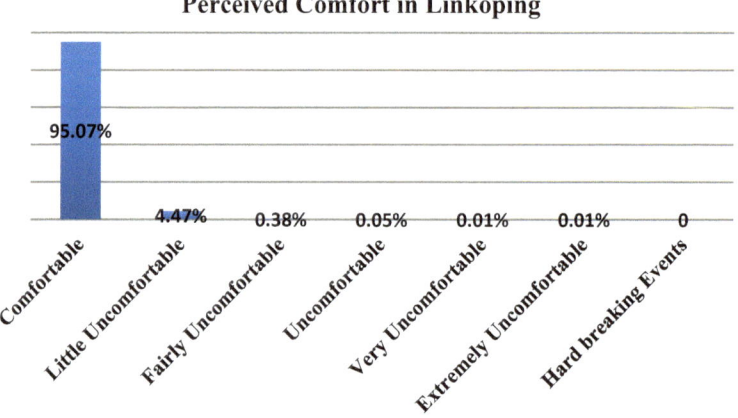

**Fig. 6** Perceived Comfort in Linköping. *Source* Authors' own pictures

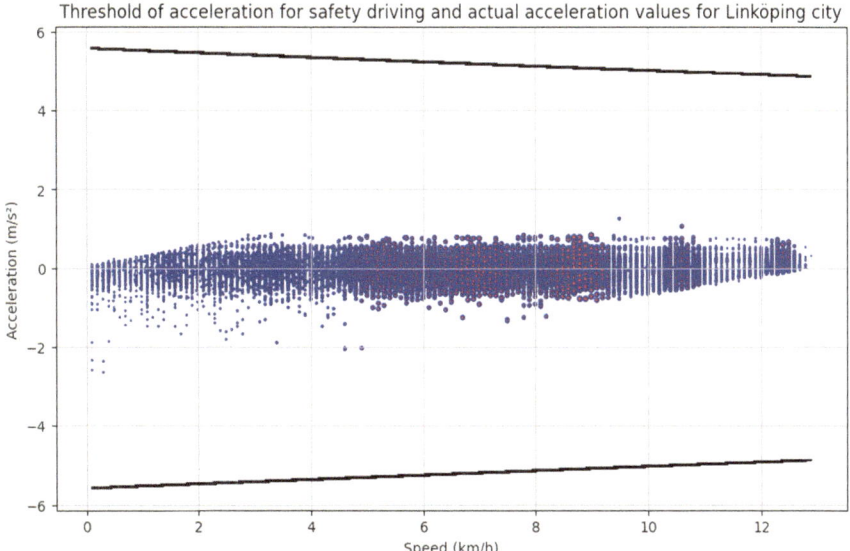

**Fig. 7** Safety acceleration profile for Linköping. *Source* Authors' own pictures

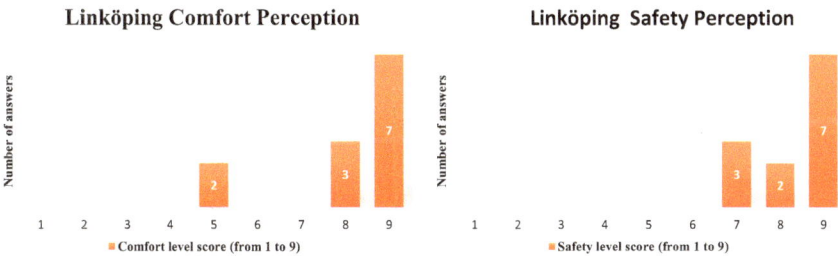

**Fig. 8** Ratings from passengers in Linköping, related to comfort (left) and safety (right) perception. *Source* Authors' own pictures

site in comparison with Graz is also expressed by the actual comfort rating, since the mean value and the variance are decreased about **15.32%** (1.05 instead of 1.24) and **85.36%** (0.06 instead of 0.41), respectively.

## 5.3 Case Study 3: Madrid

The final case study refers to the EMT bus depot site. The operation period in which both subjective and objective data is analysed herein is June 2023. The data respects to two operation days. The amount of speed and timestamp data is 15,595. The mean speed value is 6.1 km/h, the variance 5.61 km²/h², the CV 0.24, whereas the max

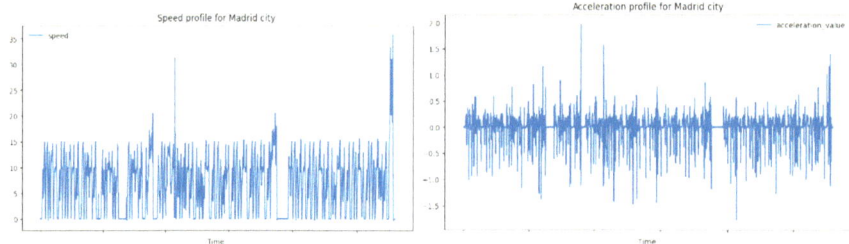

**Fig. 9** Speed (left) and acceleration (right) instances from Madrid. *Source* Athors' own pictures

speed is equal to 11.14 km/h. Figure 9 illustrates some speed and acceleration data points from 30–06–2023.

Starting with the analysis of the actual comfort, the respective percentage values per category are depicted in Fig. 10 and are "**Comfortable**" **93.51%**, "**Little Uncomfortable**" **4.53%**, "**Fairly Uncomfortable**" **1.46%**, "**Uncomfortable**" **0.78%**, "**Very Uncomfortable**" **0.054%**, "**Extremely Uncomfortable**" **0.028%** and "**Hard braking events**" **0.015%**. As concerns the alignment of each level with the comfort qualitative rating, the Madrid site has median value equal to **1**, mean value equal to **1.09** and, lastly, the variance is **0.015**.

Regarding safety levels, Fig. 11 displays the acceptable acceleration thresholds alongside the actual acceleration values for the Madrid case. It is obvious from the graph that as with the two previous pilot sites all the data points are inside the safe zone, as defined, with mean and median values equal to 1. However, the acceleration values in this case are closer to the thresholds compared to previous cases. Despite this proximity, safety is still assured for all the observed data points.

In the site of Madrid, as reflected by Fig. 12, the summary of the percentage values that belongs until the "**Fairly Uncomfortable**" class is **99.5%** (somehow familiar

### Perceived Comfort in Madrid

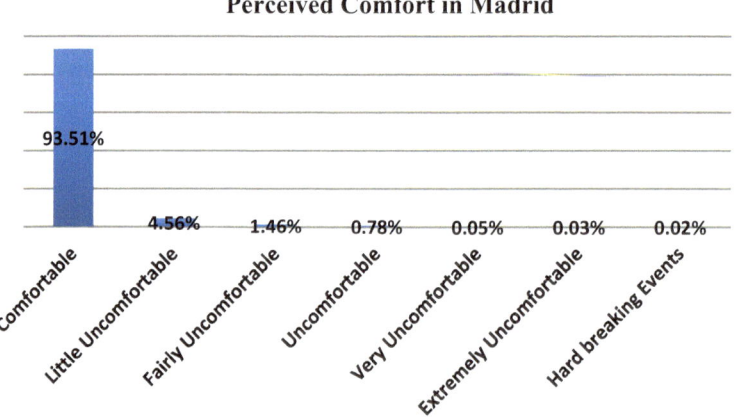

**Fig. 10** Perceived comfort in Madrid. *Source* Authors' own pictures

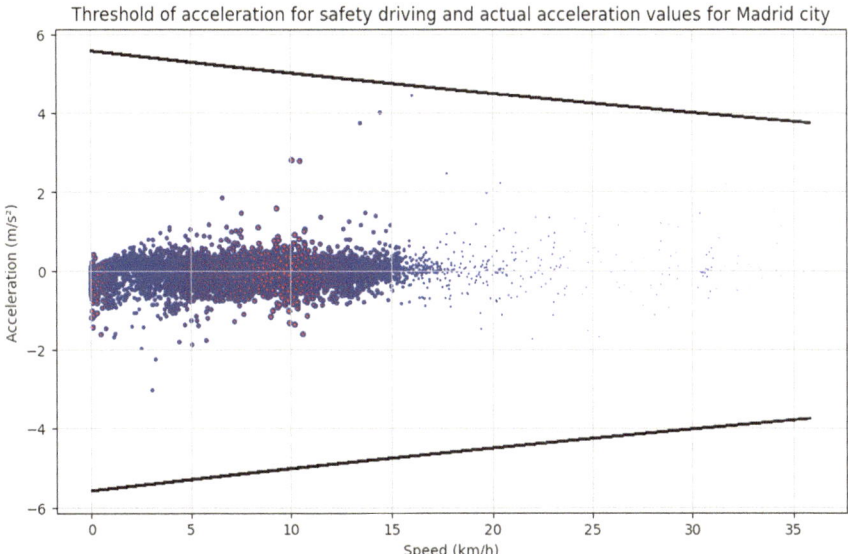

**Fig. 11** Safety acceleration profile for Madrid. *Source* Authors' own pictures

with Linköping). However, the percentage of values that belong to the higher levels (7 to 9) for comfortability and safety are **23.07%** for both metrics (3 out of 13). The lower perception levels are sourced by the existence of hard braking events that represents only the **0.015%.** Finally, the comfort rating is somehow similar to the Linköping pilot site (mean value equals to **1.09** instead of 1.05 and variance is **0.015** instead of 0.06).

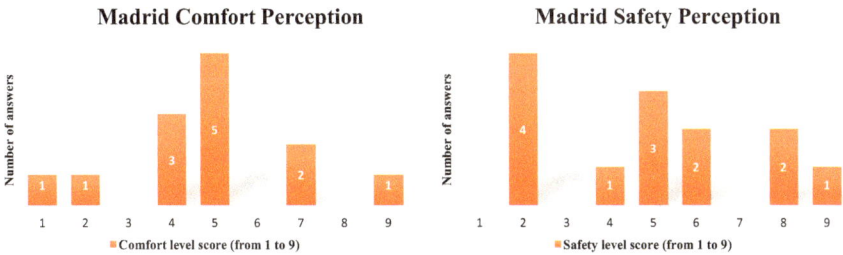

**Fig. 12** Ratings from passengers in Madrid, related to comfort (left) and safety (right) perception. *Source* Authors' own pictures

## 5.4   Correlation with Hypotheses and Discussion

Apart from the descriptive statistics presented for each of the three pilot sites (Graz, Linköping, and Madrid) in the previous subsections, in this one, the results of the statistical tests having been performed to answer the eight hypotheses, presented in the Sect. 3, are analysed.

Regarding the first four hypotheses, the results of the Pearson correlation, clearly showed a strong correlation ($\rho$ values greater than 0.8 for all the four hypotheses) between the speed profile (speed coefficient of variation) and the number of hard braking events, with the passenger comfort and safety perception levels. More specifically, as the speed coefficient of variation is getting smaller (namely, a smoother speed profile), the perception of passenger comfort and safety is getting higher ($\rho$ values of $-0.93$ and $-0.81$ for comfort and safety respectively). Similarly, as the percentage of hard braking events is getting lower, the levels of comfort and safety, as perceived by the passengers, are getting higher ($\rho = -0.99$ for comfort perception and $\rho = -0.92$ for safety perception). However, for all the aforementioned hypotheses, due to the limited number of pilot sites, the results could not be statistically significant. Nonetheless, this analysis provides a solid foundation for comparing more AV pilot sites with a broader data spectrum and testing the aforementioned hypotheses.

With respect to the last four hypotheses, related to the probable difference among the pilot sites in terms of safety, and comfort either actual or perceived, the results of the Kruskal–Wallis and Mann–Whitney U are mixed. More particularly, according to the statistical tests, there is statistically significant difference among the pilot sites with regards to the actual comfort as reflected by the four $p$-values equal to 0 corresponding to the respective statistical tests (one Kruskal–Wallis test for testing the difference in median values across all the sites, and three Mann–Whitney U tests for testing the difference in median values between all the possible pairs of the three sites). This outcome substantiates the presence of compelling evidence indicating significant discrepancies among the compared sites values in terms of actual comfort levels as translated the acceleration values from the respective literature. However, regarding the actual safety, since all the values in the three different pilot sites are inside the safe thresholds, no statistical test can be performed, and thus the hypothesis that there is no difference across the sites, in terms of actual safety cannot be rejected. Furthermore, for the hypotheses related to perceived comfort and safety, by performing the same statistical analysis, it is evident that there is a statistically significant difference in the perception of the passengers among the three pilots. For a total of eight statistical tests (for the comfort and safety perception), the only statistical test that has as a result a $p$-value different from 0 (difference in the median values of perceived comfort between Graz and Linköping). Therefore, it is more than clear that there is a major difference in the passenger perception among the pilot sites in terms of both comfort and safety (boxplots of Fig. 13 reflect graphically this difference), although that with respect to the actual safety the previous finding cannot be explained. Table 1 summarizes all the aforementioned statistical results also providing the information if a hypothesis is rejected or not (Table 2).

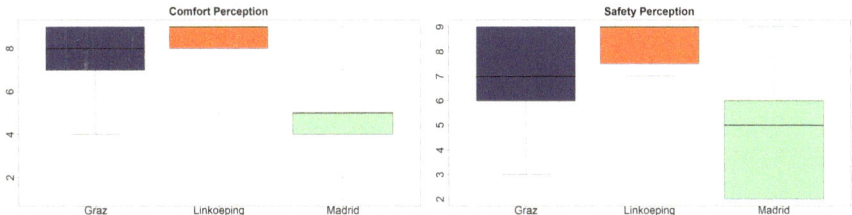

**Fig. 13** Perceived Comfort (left) and Safety (right) among the three pilot sites. *Source* Authors' own pictures

**Table 2** Hypotheses testing results

| Hypothesis | Result | Statistical significance |
|---|---|---|
| A smoother speed profile correlates with higher passenger comfort perception levels | Accepted ($\rho = -0.93$) | No ($p = 0.23$) |
| A smoother speed profile correlates with higher passenger safety perception levels | Accepted ($\rho = -0.81$) | No ($p = 0.40$) |
| A lower number of hard braking events correlates with higher passenger safety perception levels | Accepted ($\rho = -0.99$) | No ($p = 0.07$) |
| A lower number of hard braking events correlates with higher safety perception levels | Accepted ($\rho = -0.92$) | No ($p = 0.25$) |
| There is no statistical difference regarding the actual comfort among the pilot sites<br>Graz-Linköping<br>Graz-Madrid<br>Linköping-Madrid | Rejected ($x^2 = 11017$)<br>Rejected ($W = 775224562$)<br>Rejected ($W = 160245390$)<br>Rejected (W = 567425847) | Yes ($p = 0$)<br>Yes ($p = 0$)<br>Yes ($p = 0$)<br>Yes ($p = 0$) |
| There is no statistical difference regarding the actual safety among the pilot sites | Accepted | – |
| There is no statistical difference regarding the comfort perception among the pilot sites<br>Graz- Linköping<br>Graz-Madrid<br>Linköping-Madrid | Rejected ($x^2 = 18.79$)<br>Accepted ($W = 221.5$)<br>Rejected ($W = 503.5$)<br>Rejected ($W = 17.5$) | Yes ($p = 0$)<br>No ($p = 0.31$)<br>Yes ($p = 0$)<br>Yes ($p = 0$) |
| There is no statistical difference regarding the safety perception among the pilot sites<br>Graz-Linköping<br>Graz-Madrid<br>Linköping-Madrid | Rejected ($x^2 = 16.33$)<br>Rejected ($W = 139.5$)<br>Rejected ($W = 442$)<br>Rejected ($W = 16.5$) | Yes ($p = 0$)<br>Yes ($p = 0$)<br>Yes ($p = 0$)<br>Yes ($p = 0$) |

# 6   Final Conclusion and Additional Insights

In this study, we conducted the first comprehensive correlation analysis between CAV performance data and subjective measures across three European pilot sites. These sites are mature, operating in entirely different environmental conditions. The subjective data from all sites indicate that they offer completely secure transportation and satisfactory comfort.

Passenger perceptions of comfort and safety can be explained by the performance data analysis, although not perfectly. The analysis has shown that a smooth speed profile, resulting in lower acceleration values, enhances the perception of comfort and safety. Additionally, a lower number of hard braking events, defined as acceleration values below $-3$ m/s$^2$, is the second factor contributing to higher comfort and safety perceptions.

These findings were observed in sites that significantly differ from each other, as evidenced by the investigation of the respective hypotheses. This suggests that the operation of autonomous fleets can vary widely depending on the technology used, the area of operation, and the environmental conditions. However, this diversity strengthens the robustness of the analysis. By establishing key statistical tools and thresholds, this analysis could be extended to other European CCAM projects, such as ULTIMO.[3] This would provide a unified framework for assessing comfort and safety levels, helping to address objective vulnerabilities and focus on subjective issues. Ultimately, this could enhance trust in both current and future AV systems.

**Acknowledgements**  The work presented in this manuscript has been funded by the SHOW project which has been made possible by funding from the European Union's Horizon 2020 Research and Innovation Programme under Grant Agreement no. 875530.

# References

Anund A, Ludovic R, Caroleo B, Hardestam H, Dahlman A, Skogsmo I et al (2022) Lessons learned from setting up a demonstration site with autonomous shuttle operation–based on experience from three cities in Europe. J Urban Mobility 2:100021

Armstrong RA (2019) Should Pearson's correlation coefficient be avoided? Ophthalmic Physiol Opt 39(5):316–327

Asua E, Gutierrez-Zaballa J, Mata-Carballeira O, Ruiz JA, Campo I (2022) Analysis of the motion sickness and the lack of comfort in car passengers. Appl Sci 12(8):3717

Atarod M (2021) An evaluation of occupant dynamics during moderate-to-high speed side impacts. Proc Inst Mech Eng Part H: J Eng Med 235(5):546–565

Eboli L, Mazzulla G, Pungillo G (2016) Combining speed and acceleration to define car users' safe or unsafe driving behaviour. Transp Res Part c: Emerg Technol 68:113–125

European Committee for Standardisation (2020) Transportation-logistics and services-public passenger transport-service quality definition, targeting and measurement. Standard DIN EN 13816

---

[3] https://www.ultimo-he.eu/.

Friman M, Lättman K, Olsson LE (2020) Public transport quality, safety, and perceived accessibility. Sustainability 12(9):3563

Graz site (2024) SHOW website. https://show-project.eu/mega-sites-austria/

İmre Ş, Çelebi D (2017) Measuring comfort in public transport: a case study for İstanbul. Transp Res Procedia 25:2441–2449

Jalilibal Z, Amiri A, Castagliola P, Khoo MB (2021) Monitoring the coefficient of variation: a literature review. Comput Ind Eng 161:107600

Kumar V (2018) Modelling and simulation of a passenger car for comfort evaluation. Int J Res Appl Sci Eng Technol 6

Linköping site (2024) SHOW website. https://show-project.eu/mega-sites-sweden/

MacFarland TW, Yates JM, MacFarland TW, Yates JM (2016) Mann–whitney u test. Introduction to nonparametric statistics for the biological sciences using R, pp 103–132

Madrid site (2024) SHOW website. https://show-project.eu/mega-sites-madrid/

McKight PE, Najab J (2010) Kruskal-wallis test. The corsini encyclopedia of psychology, pp 1–1

Moody J, Bailey N, Zhao J (2020) Public perceptions of autonomous vehicle safety: an international comparison. Saf Sci 121:634–650

Netigate (2024). SHOW website. https://show-project.eu/citizens-engagement/

Papadopoulos A, Sersemis A, Spanos G, Lalas A, Liaskos C, Votis K, Tzovaras D (2024) Lightweight accident detection model for autonomous fleets based on GPS data. Transp Res Procedia 78:16–23

SHOW (2020) D9.1: Evaluation framework. Deliverable of the Horizon 2020 SHOW project, Grant Agreement No. 875530. https://show-project.eu/wp-content/uploads/2021/04/SHOW-WP09-D-UIP-002-01_-_SHOW_D9.1_Evaluation_Framework_SUBMITTED.pdf

SHOW Project. (2024, May 9). SHOW. https://show-project.eu/

Toloudis D, Spanos G, Angelis L (2016) Associating the severity of vulnerabilities with their description. In: Advanced information systems engineering workshops: CAiSE 2016 international workshops. Ljubljana, Slovenia, Proceedings 28. Springer, Berlin, pp 231–242

# Innovations and Collaborations

# Driving the Future: Unveiling Innovative Business Models for Shared Automated Mobility Services

Jaâfar Berrada, S. M. Hassan Mahdavi, Romina Quaranta,
Paola Rodríguez, Victor Ferran, and Jenny Weidenauer

**Abstract** Automated vehicles are becoming more and more likely as they offer promising solutions to improve urban mobility: dispatching vehicles on roads to minimize congestion, reducing accidents, increasing savings of travel time, improving the transit level of service and reducing operating costs of public modes, thus limiting public subsidies. However, these potential positive externalities are not a sufficient condition for their deployment's success. A viable Business Model has to be developed. This paper aims to provide a better understanding of the emerging Business Models in shared Cooperative, Connected and Automated Mobility (CCAM), through analysing the experiences and lessons learned from two pilot sites within the SHOW project: Les Mureaux (France) and Monheim am Rhein (Germany). In particular, a series of interviews have been conducted with mobility experts from both sites and conditions to achieve the viability whilst, later, the scalability of Business Models were discussed. Results revealed that to make automated transport services successful, several adjustments to traditional Business Models are necessary in terms

J. Berrada (✉) · S. M. Hassan Mahdavi
Department of New Solutions of Mobility Services and Shared Energy, VEDECOM, 23 Bis Allée
Des Marronniers, 78000 Versailles, France
e-mail: jaafar.berrada@vedecom.fr

S. M. Hassan Mahdavi
e-mail: hassan.mahdavi@vedecom.fr

R. Quaranta
Department Automated Driving, T-Systems International GmbH, Marsplatz 4, 80335 Munich,
Germany
e-mail: romina.quaranta@t-systems.com

P. Rodríguez · V. Ferran
Bax and Company, C/ de Casp, 118, 120, L'Eixample, 08013 Barcelona, Spain
e-mail: v.ferran@baxcompany.com

J. Weidenauer
IESTA – Institute for Advanced Energy Systems & Transport Applications, Strasserhofweg 9,
8045 Graz, Austria
e-mail: jenny.weidenauer@iesta.at

© The Author(s) 2025

H. Cornet and M. Gkemou (eds.), *Shared Mobility Revolution*, Lecture Notes
in Mobility, https://doi.org/10.1007/978-3-031-71793-2_12

of costs structures, required resources, and key partnerships. The viability and scalability are also sensitive to the costs of vehicles and supervision as well as to the maturity of automation technology.

**Keywords** Shared automated vehicles · Mobility services · Business models · Viability · Scalability

# 1  Introduction

## *1.1  Context*

In recent years, there has been a growing body of literature addressing various aspects of Automated Vehicles (AVs). Researchers have made significant contributions in areas such as common evaluation methodologies (Innamaa et al. 2023; Barnard and Sami 2023), impact assessment, simulation (Berrada and Leurent 2017), user acceptance (e.g. Bansal et al. 2016, KPMG 2019), and regulatory frameworks (e.g. Synced 2018, BakerMcKenzie 2018). Some of these studies have been based on real life demonstrations, either within the framework of national initiatives like SAFESTREAM (SAFESTREAM) and SAM (SAM 2019–2024), or as part of European projects such as AVENUE (AVENUE). These studies have shed light on the technical capabilities of AVs, examined user attitudes and preferences, and investigated the legal and policy implications surrounding their deployment.

Despite this progress, the evaluation of Business Models for AVs has received comparatively less attention. While researchers have delved into different dimensions of Business Models, including their definition, taxonomy, constituent elements, ontology, and design tools, the process of evaluating Business Models—particularly before their introduction in the market—remains an underexplored area. Centred around Business Models for Shared Automated Vehicles (SAVs), this chapter provides an in-depth evaluation of two SHOW (SHOW 2020) pilot sites for which data collection and discussion are among the most advanced and have reached an advanced state of service deployment: (1) Monheim am Rhein (Germany), featuring a fixed-line service on public streets and the narrow city centre. (2) Les Mureaux (France), featuring automated public transport in private site without an onboard supervisor.

## 1.2   Boundary Conditions

The dynamics of Business Models are shaped by several factors, including techno-logical advancements, regulatory frameworks, consumer preferences, and environ-mental concerns. However, amidst this complex landscape, certain boundary condi-tions serve as critical pillars upon which successful Business Models for SAVs must be built:

- **Electric propulsion as the norm**: In the context of SAVs, all vehicles are assumed to be electrically powered. This foundational premise not only aligns with global trends towards sustainable transportation but also underscores the imperative for Public Transport Operators (PTOs) to prioritize eco-friendly solutions in their fleets.
- **Full autonomy as the end goal**: The objective for PTOs is to achieve fully automated, driverless operations not necessarily by deploying their whole fleet and transforming their full operation respectively, but by selecting specific use cases that would contribute in a meaningful manner to a holistically sustainable mobility system. While the timeline for achieving this goal may vary depending on technological maturity and regulatory constraints, automation is crucial to fill gaps in the mobility ecosystem which would be too expensive otherwise to support. Such mobility complements would encompass feeder services, off-peak-hours-/on demand services, or services accommodating rural areas or areas with a special demography (e.g., kids, elderly).
- **Integration into the Public Transportation Ecosystem**: Shared Automated Electric Vehicles (SAEVs) are not standalone entities but integral components of the broader public transportation ecosystem. Business Models for SAEVs must, therefore, emphasize and consider interoperability and effective integration with existing modes of transportation, including buses, trains, and micro-mobility.

## 1.3   Research Questions

To gain a better understanding of the emerging Business Models in Cooperative, Connected and Automated Mobility (CCAM), this article delves into the experiences and lessons learned from the pilot sites within the SHOW project. SHOW represents a significant initiative aimed at showcasing the practical implementation and viability of shared CCAM technologies and services. The following research questions will be answered for the two selected pilot sites of the project in Les Mureaux (France) and Monheim am Rhein (Germany):

- How do established mobility services in public transportation differ from envisaged operations which integrate automation?
- Which are those Business Models elements that turn them viable for SAVs?
- What are the factors that need to evolve, to enable currently not (or not fully) viable Business Models turn viable in the future?

The following Sect. 2 will discuss the definitions of the key elements. Section 3 explains the exploration methodology. Section 4 examines the current Business Models, including insights on Business Model pathway (4.1), the viability (4.2), and scalability of the Business Models (4.3). Finally, Sect. 5 provides a summary of findings and conclusions.

## 2 Definitions

### 2.1 Definition of Business Models

Business Models describe the methods by which an organization or sector seeks to create and capture value. This includes strategies for revenue generation, value proposition, competitive positioning, and customer engagement (Bouwman et al. 2008). Effective Business Models in CCAM must navigate the complex interplay of rapidly evolving technologies, shifting regulatory landscape, and changing consumer preferences, all while striving to achieve sustainability and profitability.

The **key elements** to evaluate the Business Models in SAV, can include four different perspectives: (i) the users' perspective, through the analysis of acceptability; (ii) the service provider's perspective, in terms of efficiency and cost estimation; (iii) the quality of service, requiring the treatment and analysis of collected data; and (iv) the society's perspective, in terms of environmental impacts, safety and quality of life. Nielsen (2014) The chapter focuses on the service provider's perspective, namely the PTOs, and on the service setup and provision as well as the differences to be expected in terms of costs and savings between established services and the automation of fleets.

### 2.2 Definition of Viability

A viable Business Model is essential for the success and the long-term survival of any project or service (Magretta 2002). A Business Model becomes viable when all stakeholders can derive value from it, fostering their engagement and commitment (Chesbrough and Rosenbloom 2002). The easiest way to do this is to assess the profitability of each stakeholder. Furthermore, for non-profit-driven stakeholders, their value capture is assessed qualitatively in terms of benefits generated (D'Souza et al. 2014; Gordijn and Akkermans 2003). For a Business Model to be viable it also must be technologically viable (Kraussl 2011). A Business Model is technologically viable when an acceptable technological solution enables the provision of the envisioned service. In conclusion, a Business Model is viable when it is viable in terms of value and technology both.

## 2.3  Definition of Scalability

Scalability is part of the Business Model exploitation; it describes the ability of a system to adapt easily to increased workload or demand. Business Model scalability is seen, thus, as its ability to benefit from economies of scale. For instance, the ratio between the costs/efforts and the revenues/benefits of putting a new service in place as a proxy to determine a scalability potential can be used.

According to Chesbrough et al. (2006), the Business Model dimension evaluates an automated vehicle operator's ability to create a scalable business that will result in sustainable profitability. Identifying and assessing scalable Business Models is complex, especially in consumer transportation. For example, robo-taxis offering rides to and from major airports may combine price/mile with in-vehicle advertising, allowing them to charge lower prices while still showing higher profits. This can be applied to the present cases like daily commute rides such as on-site private demand and last-mile services.

To evaluate the scalability of a Business Model it has been compiled the following set of criteria based on the literature (Chesbrough et al. 2006; Simoudis 2019): (i) define the growth vision, identifying where the business will be in a specified timeframe; (ii) analyse the current capacity and identify bottlenecks, by understanding current operational limits (technology, human resources, capital, regulatory issues, infrastructure); (iii) analyse how the variable versus fixed costs could behave as the business expands. If costs increase linearly with growth, scalability will be challenging; (iv) assess if the current technology stack (e.g., sensors, etc.) and resource availability (e.g., more vehicles, skilled technicians) support rapid scaling, and can be easily obtained as the business grows; (v) evaluate if operational processes are streamlined and can be replicated easily in higher volumes; and (vi) study potential markets or demographics not currently being served by revising if there are robust systems in place for collecting and analysing user feedback as the Business Model scales.

In this article, the aim is to assess the scalability of the Business Models evaluating how the deployment of SAV in one French and one German pilot site can be adapted and expanded to meet the demands of a larger market while maintaining their effectiveness sustainability and address its challenges before they magnify in the full-scale expansion.

## 3  Exploration Method

### 3.1  Description of Applications Cases

The two Business Models discussed are the ones deployed at Monheim am Rhein (Germany) and Les Mureaux (France) pilot sites. The PTOs "Bahnen der Stadt Monheim" (BSM) and "Transdev" are respectively the operators of the services.

In Monheim, the service which can be seen in Fig. 1 started in 2020 and since then operates as an addition to the traditional bus services in the city. The route for the automated shuttles is a 1.7 km fixed-line and passes the urban areas featuring a health campus, residencies for elders and the old city part, passing through an old tower in the pedestrian-only zone. Therefore, the target group for the automated shuttles is mostly elderly people and young families, but the service attracts a lot of tourists as well. More than 30 specially trained safety operators work to provide this special service.

In Les Mureaux, the remotely operated shuttle service is located on the private site of ArianeGroup, a French aerospace company. The service benefits the employees and reduces travel time onsite and to the site's gates. The roads on the site include two-way sections, multiple pedestrian crossings, parking areas, roundabouts, areas with dense buildings and vegetation, and open areas (plain at the end of the aerodrome runway). Multiple stops have been established, and the shuttle operates with scheduled stops and regular departures. The transition to operate without an onboard operator was completed on October 1st, 2022. Services without onboard operators for regular routes cover distances of 2.2 and 2.6 km.

## 3.2 Analysis Framework

To assess the viability and scalability of various Business Models, according to the definitions above, a series of expert interviews with PTOs from both sites Les Mureaux and Monheim were conducted.

Subsequently, two distinct series of online structured interviews with local transport operators (PTOs) of several SHOW pilot sites were conducted. In particular:

**Table 1** Structure and rounds of the interviews

| Interview | Objectives | Structure |
|---|---|---|
| First round | Understanding the business model and initial learnings | Section 1: Presentation of city case: motivation, strategy<br>Section 2: Main assumptions of the business model<br>Section 3: Business model<br>Section 4: Lessons learned<br>Section 5: Future plans |
| Second round | Exploring viability and scalability conditions | Section 1: Presentation of city case: business model in place and change of business model in the future<br>Section 2: Viability measurement and conditions<br>Section 3: Scalability and replicability conditions |

- The first was a session aimed at understanding the intricacies of the Business Model (value proposition, key partners, channels, cost structure, etc.) and the initial learnings from the field trials.
- The second round constitutes another session with the same PTOs to explore deeper into their perception of the critical aspects of viability for various stakeholders, as well as the factors influencing the scalability of the Business Models.

Table 1 presents the structure of the interviews.

# 4 Results

## 4.1 Business Models

The advent of automation in transportation presents both challenges and opportunities for businesses operating in this sector. By examining the state-of-art and further plans of the pilot sites previously presented, light can be shed on the feasibility, scalability, and potential hurdles associated with integrating automated transport services into existing Business Models of the PTOs, meaning integrating the services into the established fleets and their operation as an addition to the mobility portfolio. In Figs. 2 and 3 the Business Models of both pilot sites are concluded, showing the current state-of-art of their operation, tasks and goals. Nevertheless, during the expert interviews, it became clear that all operations possible with the current technology and regulations are just the beginning.

To make automated transport services successful, several adjustments to traditional Business Models are necessary. Traditional transport services often allocate significant resources to human labour, maintenance, and infrastructure. Shifting towards automated transport requires reallocating investments towards technology development, infrastructure upgrades to support automation (such as sensors,

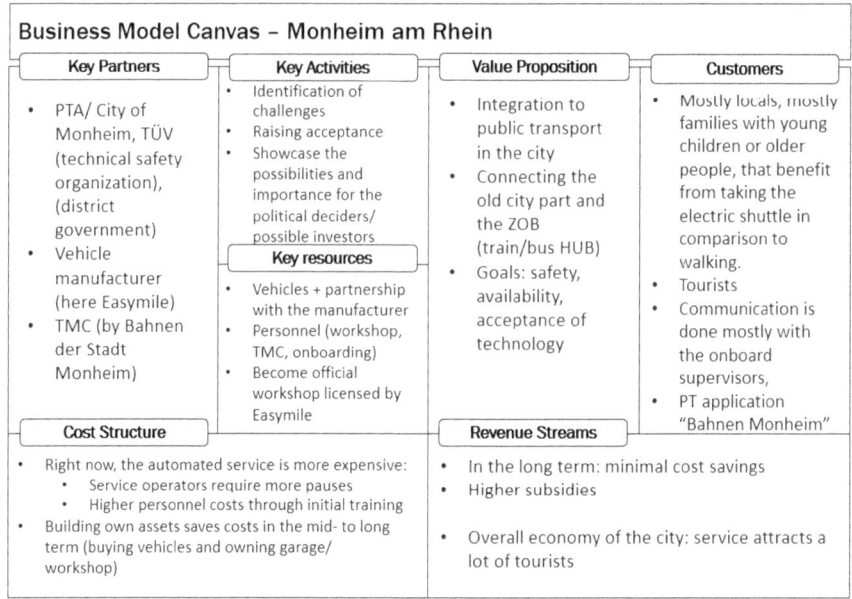

**Fig. 2** Business model canvas of Monheim am Rhein

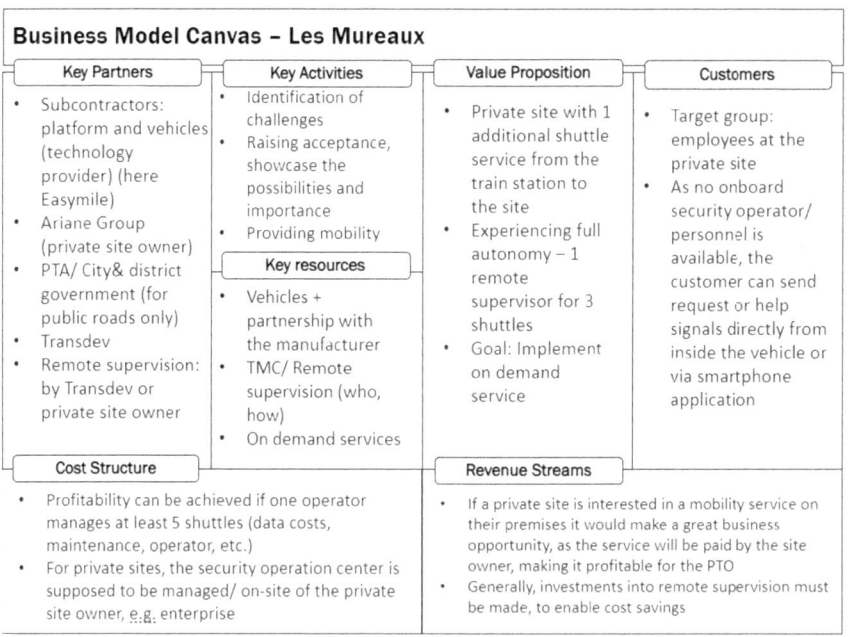

**Fig. 3** Business model canvas of Les Mureaux

communication systems, and charging stations for electric vehicles), and cybersecurity measures such as intrusion detection systems. Rather than solely relying on ticket sales, revenue might come from subscription models, advertising within vehicles, data monetization, or partnerships with other businesses for last-mile deliveries or shuttle services.

Collaboration with stakeholders across the transportation ecosystem, including technology providers, infrastructure developers, urban planners, and public agencies, is crucial. Forming strategic partnerships can help businesses access expertise, resources, and funding opportunities to scale automated transport services effectively. It is important to involve decision-making bodies in the development of a service. For example, in Germany it is necessary to have every vehicle inspected by the TÜV. In the case of automated vehicles, special safety regulations apply for approval on public roads and for passenger transportation.

Demonstrating the proof of concept, reliability, safety, and performance and providing evidence of successful pilot projects can instil confidence in investors by showcasing the feasibility and scalability of automated mobility solutions and gain the trust and support of the investors.

The experts stated that the most interested partners for financing automated mobility service in Europe are the public government bodies and opportunities for public–private-partnerships (PPPs). Governments often prioritize investments in innovative transportation solutions that improve efficiency, reduce congestion, and enhance sustainability. Collaborating with private sector companies through PPPs can provide access to additional financial resources and expertise. Private partners may contribute capital investment, technology development, and operational support in exchange for a share of project revenues or other benefits. PPPs can help spread financial risk and incentivize private sector participation in automated mobility initiatives.

## 4.2   Viability of Business Models

### Viability Objectives of stakeholders

The viability of the Business Model demands diverse measurements, tailored to the interests and expectations of various stakeholders. This holds true, notably, for both the French and German contexts, encompassing two distinct Business Models—on-site private demand service and last-mile service—each with its unique set of objectives.

Within the French landscape, concerning the on-site private service, the objectives are the same for PTOs and local authorities (i.e., municipalities or cities). They strive to deliver an efficient, reliable, and indispensable mobility service to the site's employees. The pilots' purpose was precisely to validate the value for users, allowing PTOs to explore technological maturity and Public Transport Authorities (PTAs) to assess safety.

On the German front, the Business Model's viability hinges on creating a seamless service chain—connecting regional train stations to different city parts through buses, bikes, etc. The pilot successfully constructed a multimodal ecosystem, promoting the service to the public to assess its value for users. Simultaneously, it empowered the operator to gather usage data in diverse scenarios, accelerating technological development. For the local authority, this pilot served as a potent tool for service promotion.

### Economic viability analysis

Economic viability remains elusive with the current scale of the pilot service. Achieving this goal necessitates a reduction in vehicle costs, Hardware expenses, sensors, licenses, Application Programming Interfaces (APIs), etc. The study of Whitmore et al. (2022) estimated that automation will reduce wage costs by 60%. With the cost savings brought by remote supervision, autonomous shuttles have the potential to improve the flexibility and the accessibility of public transport (Séjournet et al. 2023). The difference between remote and on-board supervision is also location specific as higher income countries would benefit more from remote supervision (Becker et al. 2022).

According to the site's experts, scaled deployment, with at least five vehicles manageable by a single remote control, is crucial to reach the economic viability. In Les Mureaux, Transdev tested a single remote supervisor to oversee a fleet of three automated shuttles and proved that safety could be ensured. However, they argue that one supervisor for three vehicles is not economically viable and cannot cover all operating costs (e.g., data, maintenance, etc.). A recent study of Hickert et al. (2023) explored the number of supervisors required to ensure safety while considering different penetration rates of cooperative and non-cooperative automated cars (i.e., replacing personal cars). They found that one remote supervisor could operate 52 vehicles. Nevertheless, they ignored the cognitive capability (e.g., situation awareness, fatigue) of the human supervisor to manage many vehicles (McKerral et al. 2023). Coming back to the economic viability, there is currently no evidence about the minimum number of vehicles that should be supervised to be viable.

### Technological viability

The automation of vehicles is still facing major technological challenges to be viable. The maximum speed of shuttles allowing to detect obstacles and to react in emergency is under 30 km/h, which makes biking or riding e-scooters even more attractive as a last-mile solution. In addition, shuttles should analyse their environment smarter to avoid erroneous detections. Improving performances and accuracy of sensors is crucial to increase the service speed and with that the quality of service. Current developments of the new vehicle generation (Gen3) aim to overcome these challenges, by reaching higher speeds while being more reactive.

Regarding the supervision technology, there is also room for progress: control of doors and ramps, communication with on-board passengers, etc. The integration of automated vehicles into a global transportation system will in addition generate more complexities, since the two layers of supervision should cooperate: the supervision of the unit vehicle, performed by the service provider, and the supervision of all shared vehicles, including other automated cars and buses, performed by the PTO.

**Other viability conditions**

According to the experts, the acceptance of the service usage is high. On the one hand, the regulation authorities are supportive. On the other hand, capabilities of local authorities must evolve to align with the advancement of the technology, encompassing personnel training and dedicated training centres, not limited to specific projects but encompassing the broader operational framework.

**Summary of viability conditions**

In summary, realizing the viability of the two Business Models demands the deployment of a service that attains commercial-level performance, rivalling the efficiency of current mobility services, primarily sourced by public authorities at local, regional and/ or national levels, and seamlessly integrates into a comprehensive multimodal landscape (Table 2).

**Table 2**  Conditions of viability

| Viability | Conditions | Comment |
| --- | --- | --- |
| Economic viability | Decrease of vehicles costs (sensors, LIDARs, hardware, etc.) | However, more advanced sensors and LIDARs are required to reach better performances |
| | One supervisor for at least five vehicles | The cognitive capabilities of the human supervisor to be evaluated<br>The minimum number of vehicles to be observed needs to be validated |
| Technology viability | Higher speeds and lower reaction time | More advanced and accurate sensors required |
| | Supervision with additional components to control the main features of the vehicle | Technology development and additional partnerships required |
| | Cooperation between different layers of supervision | Technology development and additional partnerships required |
| Other viability aspects | Acceptability is good and regulation is supportive | |
| | Skills of public authorities should evolve (digital and management skills to successfully integrate CCAM solutions into Sustainable Urban Mobility Plans) | |

## 4.3   Scalability of Business Models

The interview questions were designed to delve into the scalability of the pilot sites across various parameters. To evaluate the scalability of a Business Model, the questions focused on the following set of criteria:

- **Projected Growth Vision**: Exploring the long-term ambitions of the business concerning geographical expansion, fleet size increase, revenue targets, and other growth metrics.
- **Identifying Bottlenecks**: Pinpointing potential limitations or constraints that could impede the scalability and viability of the project, thereby allowing preventive measures to overcome them.
- **Assessing Resource Availability**: Evaluating the accessibility of necessary resources, such as additional vehicles or skilled technicians, to support the growing demands of the business.

The on-site private demand service in France contemplates expanding its services, including additional routes and operating hours in a dedicated/quasi-dedicated lane model. However, the current stage of the pilot presents challenges in providing comprehensive end-to-end services at a technical level, due to the need for additional stakeholders/operators. Technology assessment reveals that service replication is hindered by the lack of maturity in the economic model, particularly concerning maintenance costs, operator expenses, and data management. Profitability projections hinge on optimizing remote-control strategies, with each operator managing a minimum of five shuttles driving in dedicated lines to maintain current safety standard. Notably, none of the vehicles are currently homologated for commercial service, with homologation expected in 2025–2026.

A local expert recommends focusing financing efforts on established European leading companies on autonomous mobility (e.g., ADS manufacturers) and engaging key stakeholders directly, consolidating funding for fleets of at least 5–6 vehicles to drive industrialization and process improvement rather than dispersing resources across multiple projects. With increased fleet size and capacity, higher revenues are anticipated, aiming for a mid-term margin of 5–10%, although fluctuations in contract costs pose instability.

In Germany, the pilot operates a single line with automated shuttles, with no plans for expansion due to perceived high costs by citizens and lack of financial incentives. Driver shortage remains a primary constraint (UITP 2023), prompting consideration of transitioning to driverless operations and relocating to a private site for controlled testing with no driver. Addressing the current and future challenges posed by driver shortages, automated shuttles are touted as a flexible solution, catering to continuous services and journeys impractical for human drivers (Mortkowitz 2024).

German regulations pose challenges regarding driver permissions, restricting them to defined Operational Design Domains (ODDs), thereby facilitating authorization for public transport. Accelerating the deployment of safe and feasible automated vehicle technology across diverse ODDs, including low-speed shuttles for last-mile services and on-site private demand services, is crucial for testing and refining various business models.

In conclusion, proactive measures must be taken to address challenges before they escalate during full-scale expansion. This includes analysing current operational capacity, assessing technology readiness for scaling, establishing robust feedback mechanisms, and leveraging insights from other test sites within the project.

# 5  Conclusions

This article attempted to explore Business Models for services based on electrified automated vehicles. In particular, it endeavours to describe their main components and to investigate the conditions of viability and scalability. To do so, a series of interviews have been performed with PTOs from two distinct SHOW pilots: (1) Monheim am Rhein (Germany), featuring a fixed-line service on public streets and the narrow city centre, and (2) Les Mureaux (France), featuring automated public transport on private site without an onboard supervisor.

In summary, it was determined that to make automated transport services successful, several adjustments to traditional Business Models are necessary: in terms of costs structures, required resources, and key partnerships. The services based on automated vehicles will be complementary to the already established services and would add new services filling the existing mobility gaps (feeder services, special mobility needs of inhabitants, off peak hours, on demand services in low demand areas/ rural areas). In fact, the new services would be too cost intense to be provided with the "classic" setup of buses (high capacity of passengers) and automation will ultimately enable those services (through new vehicles and technology). The public government bodies should be the most interested partners for financing automated mobility services, as these solutions are expected to improve efficiency, reduce congestion, and enhance sustainability.

The viability had been explored from different perspectives: economic, technological, and others (social, regulation, etc.). The article outlines that to reach the viability, costs of vehicles must decrease substantially. Furthermore, one supervisor should manage the operation of at least five vehicles to improve the economic viability. Technological challenges to increase safety and speed still must be addressed, through improving the accuracy of sensors and the features of supervisors.

Finally, scalability emerges as a critical factor for the success of SAV Business Models. Models that demonstrate the ability to scale, meet user demands efficiently, and offer clear paths to profitability are attractive to investors. For currently non-viable Business Models to achieve viability, several key factors need to evolve, and economic models must adapt to include diversified revenue streams. The evolution of these factors will enable the transition to commercially viable, scalable models that can effectively integrate into the broader transportation ecosystem.

While we are encouraged by these exploratory results, we recognize that this chapter has its own limitations. Firstly, it is based on experts' opinions of PTOs from two pilots. Having a good knowledge of the pilot's ecosystem, they indicated also the needs and concerns of other stakeholders. However, a deeper analysis based on direct interviews with all stakeholders of the value chain would provide more accurate and solid results on viability and scalability conditions. Secondly, the viability and scalability are context specific. In order to generalize the outcomes of the article, it would be valuable to conduct interviews with other SHOW pilots and to perform a cross-country analysis. Thirdly, a qualitative analysis is proposed to identify conditions of viability and scalability. Future research will enrich these results by performing also a cost–benefit analysis.

**Acknowledgements** The research presented is a part of the SHOW project that has received funding from European Union's Horizon 2020 research and innovation programme under grant agreement no. 875530.

# References

AVENUE (n.d.) Autonomous Vehicles to Evolve to a New Urban Experience. https://h2020-ave nue.eu/

BakerMcKenzie (2018) Global driverless vehicle survey. United States

Bansal P, Kockelman KM, Singh A (2016) Assessing public opinions of and interest in new vehicle technologies: an Austin perspective. Transp Res Part C: Emerg Technol 67:1–14

Barnard Y, Sami K (2023) Methodology for field operational tests: updating the FESTA methodology for connected and automated driving pilots. Transp Res Procedia 72:2054–2061

Becker H, Becker F, Abe R, Bekhor S, Belgiawan PF, Compostella J, Frazzoli E, Fulton LM, Guggisberg Bicudo D, Murthy Gurumurthy K, Hensher DA, Joubert JW, Kockelman KM, Kröger L, Le Vine S, Malik J, Marczuk K, Ashari Nasution R, Rich J, Papu Carrone A, Shen D, Shiftan Y, Tirachini A, Wong YZ, Zhang M, Bösch PM, Axhausen KW (2022) Impact of vehicle automation and electric propulsion on production costs for mobility services worldwide. Transp Res Part a: Policy Practice 138:105–126. https://doi.org/10.1016/j.tra.2020.04.021

Berrada J, Leurent F (2017) Modeling transportation systems involving autonomous vehicles: a state of the Art. Transp Res Procedia 27:215–221

Berrada J, Mouhoubi I, Christoforou Z (2020) Factors of successful implementation and diffusion of services based on autonomous vehicles: users' acceptance and operators' profitability. Res Transp Econ 83:100902. https://doi.org/10.1016/j.retrec.2020.100902

Bouwman H, Faber E, Haaker T, Feenstra R (2008) What's Next? some thoughts and a research agenda. In: Bouwman H, De Vos H, Haaker T (eds) Mobile service innovation and business models. Springer, Berlin, pp 137–150. https://doi.org/10.1007/978-3-540-79238-3_6

Chesbrough H, Vanhaverbeke W, West J (2006) Open innovation: researching a new paradigm. Oxford University Press

Chesbrough H, Rosenbloom RS (2002) The role of the business model in capturing value from innovation: evidence from Xerox corporation's technology spin-off companies. Ind Corp Chang 11(3):529–555

de Séjournet A, Rombaut E, Vanhaverbeke L (2023) Cost analysis of autonomous shuttle services as a complement to public transport. Transp Res Procedia 72:2323-2330. ISSN 2352-1465. https://doi.org/10.1016/j.trpro.2023.11.723

D'Souza A et al (2015) A review and evaluation of business model ontologies: a viability perspective. In: Enterprise information systems: 16th international conference, ICEIS 2014. Lisbon, Portugal. Revised Selected Papers 16. Springer International Publishing

Gordijn J, Akkermans J (2003) Value-based requirements engineering: exploring innovative e-commerce ideas. Requirements Eng 8:114–134. https://doi.org/10.1007/s00766-003-0169-x

Hickert C, Li S, Wu C (Aug.2023) Cooperation for scalable supervision of autonomy in mixed traffic. IEEE Trans Rob 39(4):2751–2769. https://doi.org/10.1109/TRO.2023.3262120

Innamaa S et al (2023) Developing a common evaluation methodology for CCAM. In: 15th ITS European congress. The Game Changer, ITS

KPMG (2019) «Autonomous Vehicles Survey Report» PerkinsCoie

Kraussl Z (2011) Operationalized alignment: assessing feasibility of value constellations exploiting innovative (2011)

Magretta J (2002) Why business models matter. Harv Bus Rev 80(5):86–92

McKerral A, Pammer K, Gauld C (2023) Supervising the self-driving car: situation awareness and fatigue during highly automated driving. Accid Anal Prev 187:107068. ISSN 0001-4575. https://doi.org/10.1016/j.aap.2023.107068

Mortkowitz S (2024) Cruise challenges bump-in-the-road, Says Oxa [Online]. Available: https://www.wardsauto.com/vehicles/cruise-challenges-bump-road-says-oxa

Nielsen C (2014) Analyzing business models. In: The basics of business models. Ventus

Nordhoff S, de Winter J, Happee R, Payre W, van Arem B (2019) What impressions do users have after a ride in an automated shuttle? an interview study. Transp Res Part F Traffic Psychol Behav 63(n°):1252–269

Peng J, Huang H, Ran B, Shi Y, Zhan F (2019) Exploring the factors affecting mode choice intention of autonomous vehicle based on an extended theory of planned behavior—a case study in China. Sustainability 11(14):1155

SAFESTREAM, The shift towards fully driverless public transport. https://safestream.tech/

SAM (2019–2024) Security and acceptability for automated driving. https://www.vedecom.fr/sam-project-security-acceptability-for-automated-driving/?lang=en

SHOW(2020) SHared automation Operating models for Worldwide adoption. https://show-project.eu/

Simoudis E (2019) Requirements for growing and scaling autonomous mobility [Online]. Available: https://synapsepartners.co/six-requirements-for-growing-and-scaling-autonomous-mobility/

Synced, «Global Survey of Autonomous Vehicle Regulations,» 15 March 2018 [En ligne]. Available: https://medium.com/syncedreview/global-survey-of-autonomous-vehicle-regulations-6b8608f205f9

UITP (2023) New UITP taskforce on workforce shortage tackles transforming labour market [Online]. Available: https://www.uitp.org/news/new-uitp-taskforce-on-workforce-shortage-tackles-transforming-labour-market/

Whitmore A, Samaras C, Hendrickson CT, Matthews HS, Wong-Parodi G (2022) Integrating public transportation and shared autonomous mobility for equitable transit coverage: A cost-efficiency analysis. Transp Res Interdisc Perspect 14:100571, ISSN 2590-1982, https://doi.org/10.1016/j.trip.2022.100571

# Integrated Traffic Simulation Developer Suite for Shared Automated Mobility

**Maria G. Oikonomou, Marios Sekadakis, Apostolos Ziakopoulos, Allan Tengg, and George Yannis**

**Abstract** Within the SHOW project (GA No 875530), real-life urban demonstrations across 22 cities were conducted, exploring and validating the integration of Cooperative Connected and Automated Mobility (CCAM) in various public transport schemes. The project employs extensive traffic simulations using different tools and approaches. This chapter outlines the development of an integrated simulation suite that combines elements from the diverse simulations. The simulation suite is a web-based open access tool and offers guidelines, steps, and mathematical definitions for simulating CCAM. Designed for researchers, practitioners and even non-experts, while providing insights and results valuable to city planners. By emphasizing key findings from simulations, the application of the suite and its support for decision-making become more tangible.

**Keywords** Traffic simulation · Automated transport systems · Simulation suite · Simulation transferability · Simulation levels

M. G. Oikonomou (✉) · M. Sekadakis · A. Ziakopoulos · G. Yannis
Department of Transportation Planning and Engineering, National Technical University of Athens, 5 Heroon Polytechniou Str., 15773 Athens, Greece
e-mail: moikonomou@mail.ntua.gr

M. Sekadakis
e-mail: msekadakis@mail.ntua.gr

A. Ziakopoulos
e-mail: apziak@central.ntua.gr

G. Yannis
e-mail: geyannis@central.ntua.gr

A. Tengg
Virtual Vehicle Research GmbH, Inffeldgasse 21a, 8010 Graz, Austria
e-mail: Allan.Tengg@v2c2.at

© The Author(s) 2025
H. Cornet and M. Gkemou (eds.), *Shared Mobility Revolution*, Lecture Notes in Mobility, https://doi.org/10.1007/978-3-031-71793-2_13

# 1   Introduction

The EC Horizon 2020 project SHOW (SHOW) aims at developing and piloting shared automated mobility operating models towards Europe-wide adoption and beyond. During the project, naturalistic large scale field trials of automated vehicles of several types are taking place in 22 cities across Europe, in the form of real traffic urban (and peri-urban) shared mobility services, to investigate Automated Driving (AD) vehicles' integration in Public Transport (PT), Demand-Responsive transport (DRT), Mobility a Service (MaaS) and Logistics as a Service (LaaS) schemes. In addition, within the project, extensive traffic simulations are conducted to support the real-life demonstrations. The purpose of simulations in relation to real-life demonstrations is multifaceted. These simulation scenarios and tools are designed to enhance the outcomes of real-world pilot field experiments by offering supplementary support and insights. They effectively integrate SHOW sites into simulations to provide clarity on their value in real-life demonstrations. Key objectives of the simulations include:

- Refining driving algorithms and driving virtual test kilometers in the simulator to reapply in the field.
- Testing risky situations or advanced traffic scenarios that are difficult to reproduce in real sites.
- Providing results for impact assessment to enable better planning of CCAM in the future.
- Simulating network effects of pilot vehicles.

Three main simulation levels are examined, namely (i) street-level, (ii) city-level and (iii) simulations of Vulnerable Road Users (VRUs) in local areas. Different simulation tools and approaches have been used, optimal for each type of simulation and depending on the specific needs of different pilot sites. Consequently, a framework that combines all critical elements of the simulation approaches followed in each case such as key inputs, models and parameters (described in detail in Sect. 4) would be highly valuable. By standardising and packaging this framework through an integrated simulation suite, future users conducting CCAM simulation can be availed.

Simulating CCAM is an ongoing, challenging task, regardless of aims or scale. To handle its intricacies, currently, there is fragmentation in simulation approaches, and a plethora of tools that are used. Many overarching problems are similar, regardless of tool selection, while simulation practitioners face several common questions, with little external input to their problems.

This chapter aims to provide insights into the process followed for developing an integrated simulation suite. The simulation suite is designed to be a web-based tool that acquires a common pool of simulation data from the different simulation categories and use cases. It can be utilized for various purposes such as traffic, environmental and safety impact assessment, traffic flow analysis, and other related applications. The tool identifies the key parameters and possible methodologies to simulate automated driving and attempts to synthesize the simulations for all pilot sites of the SHOW project.

The SHOW simulation suite is formed largely as a response to this predicament, focusing on automated public transport schemes. This multi-layered, open and interactive tool will include and combine elements from diverse simulation approaches, methodologies, and tools, as well as offer guidelines, steps, and mathematical definitions for simulating automated mobility at different levels. Furthermore, the simulation suite provides transferability options and good practices where conducting such exercises to areas with little or no data.

The simulation suite is envisioned to be used by experts when solving practical problems in the field (e.g., city planning, urban policy implementation, and strategic decision-making), researchers when broadening the understanding of CCAM-oriented simulation as well as non-experts who wish to come into contact with the more technical aspects of these approaches.

This paper is organized as follows. In the next Sect. 2, the relevant research is presented, in which existing web-based tools are provided. In Sect. 3, the overall conceptualization framework about integrating the simulation suite tool for CCAM is presented. The architecture and components of the Simulation Suite are then following in Sect. 4. Finally in Sect. 5, the paper concludes with an overview of the general scope, practical applications of the tool, and the value it adds.

## 2 Relevant Global Research

Prior to the development of the SHOW Simulation Suite, extensive research was conducted to explore existing web-based tools, encompassing a range of tools, interfaces, and manuals, available to support authorities and policymakers. These resources were investigated for their utility in assisting authorities and policymakers to comprehensively understand the multifaceted impacts of CCAM on various aspects including traffic flow, traffic safety, environmental emissions, public acceptance, demand patterns, and potential shifts in travel behavior. This thorough exploration aimed to identify and exploit state-of-the-art technologies and methodologies, ensuring that the forthcoming SHOW Simulation Suite would effectively address the complex and evolving challenges of CCAM. Higher emphasis was placed in authority and policymaker tools as we expect that authorities and policymakers will be one of the main audience types for the SHOW Simulation Suite.

This initial exploration phase involved a comprehensive review of similar tools to understand their key goals, attributes and limitations. In this context, it was crucial to identify other initiatives associated with CCAM and explore whether similar tools had been developed, such as tools shared common functionalities (i.e. knowledge module) or objective and approaches relevant to the domain of CCAM. The similarity criteria encompass aspects such as the scope of functionalities, target users, underlying technologies, and intended applications. These criteria guide the comparison and assessment of the identified initiatives. To pinpoint relevant CCAM initiatives, the Automated Driving Roadmap document provided by the European Road Transport Research Advisory Council (2019) was reviewed. Additionally, the knowledge base on Connected and Automated Driving (CAD) established under the Horizon 2020 Action ARCADE (ARCADE) was consulted, encompassing a comprehensive list of all (up to date) projects related to CCAM.

These initiatives significantly shape the landscape of CCAM development up to date. Out of these ones, the LEVITATE Policy Support Tool (PST) (LEVITATE PST) and the ARCADE Knowledge Base (ARCADE) stand as the exclusive research endeavors equipped with comprehensive tools for CCAM, each one offering distinct functionalities. The SafetyCube Decision Support System (DSS) (SafetyCube Decision Support System) is also worth mentioning as a multifunctional road-safety decision tool and encyclopaedia.

- **LEVITATE PST**: The LEVITATE PST (LEVITATE PST) is the definitive decision support tool for CCAM interventions, emerging from the LEVITATE project in Horizon 2020, which stands for 'societal LEVel Impacts of connecTed and AutomaTed vehiclEs'. As an open-access, web-based system, it offers stakeholders access to methodologies, results, bibliography, documentation, CCAM guidelines, and a Decision Support System with forecasting and backcasting capabilities. Tailored to meet stakeholder needs, the PST serves as a comprehensive resource for informed decision-making in CCAM. More details can be found in Ziakopoulos et al. (2022).
- **ARCADE & FAME CAD Knowledge Base**: CAD Knowledge Base is a pivotal repository developed under the ARCADE (Aligning Research & Innovation for Connected and Automated Driving in Europe) project (ARCADE) and its extension FAME (Framework for coordination of Automated Mobility in Europe). It consolidates CAD information, spanning projects, regulations, policies, strategies, action plans, guidelines, and evaluation methodologies. More details can be found in the official website (ARCADE).
- **SafetyCube DSS**: The SafetyCube DSS (SafetyCube Decision Support System) is the main product of the SafetyCube project within the respective Horizon 2020 Programme. It aids evidence-based policy-making, offering interactive information on road accident risk factors and safety countermeasures. The corresponding knowledge module synthesizes documents on risks, impacts, injuries, and accident scenarios. The estimator module calculates the Economic Efficiency Evaluation (E3) of safety measures, incorporating effectiveness percentages and costs for cost–benefit analysis. More details can be found in Martensen et al. (2019).

While existing initiatives such as LEVITATE, ARCADE and FAME, and SafetyCube contribute significantly to the landscape of transportation research, there remains a notable gap in the availability of comprehensive simulation tools specifically tailored to address the challenges of automated mobility. Among the reviewed tools, only some are directly relevant to the domain of automated mobility.

Specifically, the LEVITATE PST that serves as a decision support tool for CAV interventions, lacks the intuitive and multi-level functionality provided by the SHOW Simulation Suite, which not only guides simulations but also includes examples and complementary data files essential for simulation. Similarly, while ARCADE and FAME's CAD Knowledge Base consolidates information from various projects and initiatives related to CAD, it lacks a dedicated tool for assessing the impact of automated mobility, a key objective of the SHOW Simulation Suite. Finally, the SafetyCube DSS aids evidence-based policy-making by providing information on road accident risk factors and safety countermeasures. However, its focus on road safety does not directly address the complexities of automated mobility.

The SHOW Simulation Suite fills the gap in available simulation tools by offering a comprehensive platform designed specifically for automated mobility. Its interactive features provide users with insights into the complexities of simulations conducted in the SHOW project, showcasing the seamless interplay between diverse simulations and addressing the multifaceted impacts of automated mobility on various aspects of transportation. SHOW pioneers a simulation suite that facilitates robust simulations and serves as an educational tool, providing invaluable insights and guidance to researchers at various career stages. The interactive material within the suite illuminates the intricacies of different simulation software employed, emphasizing a cohesive approach to address the complexities of automated mobility. Additionally, the suite goes beyond its role as a research tool, aiming to disseminate simulation results widely and make them accessible to researchers, policymakers, city planners, and practitioners. By providing an intuitive platform for exploring these results, SHOW contributes to a greater understanding of automated mobility and its implications for diverse stakeholders.

## 3   Conceptualization of an Integrated Simulation Suite Tool for CCAM

Within SHOW, different simulation tools and approaches ideal for each occasion were used, depending on the different pilot site needs and partner expertise. Pursuing to exploit this work and knowledge beyond the scopes of the project, the value of integrating it to a concrete knowledge base and tool to be used and further exploited by the broader CCAM community was conceptualized. To allow this, a methodology to acquire a common pool of simulation data from the different scenarios and use cases, to identify the key parameters and possible methodologies on automated driving simulation and to synthesize the simulations conducted for different test sites of the project was required. As such, the development of the SHOW Simulation Suite has adopted an incorporated approach.

The main idea of the simulation suite is to combine the knowledge gained in simulating automated mobility and integrate the fundamental aspects of this procedure at its optimal level. This is being accomplished by the development of a web-based front-end tool that provides guidelines about simulation of automated driving and includes (i) useful insights for simulating automated mobility across the different pilot sites (i.e. simulation tools, required input data, parameters for modelling AD vehicles, behaviour models, etc.), (ii) simulation transferability approaches, (iii) interplay between simulations application and outcomes, and (iv) a library including visualised instructions in the used simulation modelling software and tools. Therefore, the simulation suite is designed to be a tool that constitutes a common pool of simulation outcomes, designs, and specifications for different types of simulation (microscopic traffic simulations, macroscopic traffic modelling) and use cases in the CCAM domain. By defining the type of results provided and emphasizing key findings, the application of the simulation suite and its support for decision-making become more tangible.

For this reason, the suite is designed to:

- Offer comprehensive information about the possible tools and layers available and used within SHOW, as well as suitable scenarios, such as varying percentages of automated vehicles penetration, and their potential impacts on traffic flow dynamics.
- Provide clear guidelines to assist users in defining their desired use case or study area for studying automated mobility and to further give directions for its simulation. A use case refers to a specific situation in which automated mobility could be utilized. It outlines the purpose, goals, and context of employing AD technology. Use cases can vary based on factors such as location, demographic, infrastructure, and purpose.
- Incorporate a detailed explanation of the mathematical principles that govern the simulation environments, enhancing user understanding and ability to utilize the tools effectively.

## 4 Structure of the Simulation Suite

### 4.1 Simulation Categories and Cases

This subsection gives an overview of the categories of the simulation sites within the context of SHOW. This classification was conducted based on the respective real-life pilot activities of SHOW, the corresponding test site characteristics and their associated needs for simulation. All the simulation efforts were split into three dedicated categories as follows:

- **Simulations of VRUs in local areas**: This category covers applications focusing on VRUs and shared spaces. The scope of these simulations is the safety of all VRUs in the vicinity of AD vehicles, such as pedestrians, cyclists, etc. Passengers on board in vehicles are considered out of the scope of this cluster of simulations. For most SHOW pilot cases, the bus stop is the situation in which an AD vehicle comes close to VRUs consisting mainly of possible passengers. Most contributing experts considered a bus stop as an important element from the point of view of an ego vehicle serving this bus stop. This means that a bus stop is an essential part within the simulations street level simulations, where safety and pedestrian aspects need to be included with the focus always lying on one vehicle. In some cases, the scope is to study the interactions of AD vehicles with pedestrians and not necessarily passengers. This is especially important in bus terminals with a higher number of pedestrians, where AD vehicles need to pass through. In this case, also the focus may also include analyzing dwelling time (how long passengers would need to egress and board a vehicle) and its influence on traffic flow and time management, ensuring comprehensive understanding and optimization of urban mobility dynamics.

- **Street level simulations**: In street level simulations, both operation routes and served stops are pre-defined and fixed. However, various parameters can be adjusted to investigate interactions between different types of road users and explore AD-logic and safety issues. These parameters include factors such as vehicle speeds, acceleration rates, reaction times, and behavior models for CCAM, pedestrians and cyclists. Microscopic traffic simulation techniques are employed, either independently or coupled with other simulation-related tools, to facilitate this analysis. Furthermore, the respective focus is put rather on the test site level than on the whole city/region level. Accordingly, fluctuations/shifts in transport mode choice is not the primary focus here.
- **City level simulations**: In this category, automated shuttles are simulated at city level using DRT services. The city level scenario includes both DRT on fixed routes as well as station-based DRT services with fixed stations but without fixed routes for door-to-door services. A critical difference to simulations of VRUs is that simulation herein does not only include the microscopic level at different degrees of detail, but the macroscopic level as well, aiming at providing region or city-wide results on the traffic, environmental and safety impact of AD vehicles for different implementations of automated DRT services. The extension from local to city wide simulations enables the DRT simulations to address additional Key Performance Indicators (KPIs) like the modal split changes (i.e. the share of each mode choice in number of trips or distance travelled) and others (i.e. reduction in travel time for passengers using DRT services, changes in kilometers traveled due to shifts in travel behaviour, etc.) due to the introduction of automated DRT services, compared with above categories (street and city level simulations).

### 4.1.1 Study Cases

The simulation categories and use cases are the initial aspects of the first layer (Simulating Automated Mobility) of the simulation suite, as shown in Fig. 1. Firstly, this aspect concerns the three simulation categories that are examined within the simulation efforts of the SHOW project namely simulations of VRUs in local areas, Street-level and City-level simulations. This implies that as a first step, the SHOW Simulation Suite web-based tool users will be able to choose the category that they are interested in. Secondly, the user will be prompted by the tool to choose among specific study areas ("use cases") of the eleven (11) pilot sites i.e., Brainport, Graz,

Karlsruhe, Klagenfurt—Carinthia, Linköping, Madrid, Monheim am Rhein, Rome, Salzburg, Tampere and Trikala. The three simulation categories matched with the corresponding pilot based use cases are presented in Table 1.

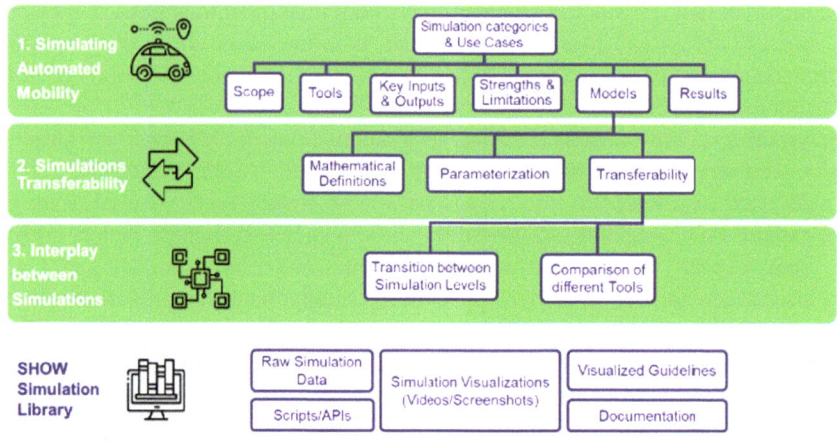

**Fig. 1** Components of SHOW simulation suite. *Source* Authors' own picture

**Table 1** Simulation categories and corresponding pilot sites

| SHOW Pilot sites | Simulation tools | Simulations of VRUs | Street level simulations | City level simulations |
|---|---|---|---|---|
| Brainport | VISSIM, New Mobility Modeller, Urban Strategy, SIL Simulator | | | X |
| Graz | ROS, Autoware simulator, SUMO | X | X | |
| Karlsruhe | ROS, SUMO, Menge, CARLA, Gazebo | X | X | |
| Klagenfurt—Carinthia | SUMO | | | X |
| Linköping | SUMO | X | X | |
| Madrid | AIMSUM | X | X | |
| Monheim am Rhein | SUMO | X | X | |
| Rome | AnyLogic, TransCAD | | X | |
| Salzburg | MATSim, SUMO | | | X |
| Tampere | AVSS | X | X | |
| Trikala | SUMO | | X | |

## 4.2  Architecture of Simulation Suite

Within the three aforementioned simulation categories and the 11 pilot sites operating CCAM, different simulation scenarios and use cases were investigated. Moreover, it is important to note that the term "simulated scenario" refers to the various inter-use cases investigated for each pilot site. For example, a pilot site integrating a point-to-point shuttle may simulate different operational speeds for the shuttle, thereby running corresponding scenarios for each speed variant.

In order for this to succeed, the SHOW simulation suite is designed to compose three different layers, as shown in Fig. 1, namely (i) Simulating Automated Mobility, (ii) Simulation Transferability, (iii) Interplay between Simulations as well as the SHOW simulation Library, which is the repository of fundamental information regarding simulating automated mobility of each layer.

The structure of the Simulation Suite website comprises multiple sections. Positioned at the top of the home page, users are able to discover a fixed navigation bar facilitating easy access to pivotal sections, including About, Street Level Simulations, City Level Simulations, VRU Simulations, and the SHOW project. Additionally, nested within the sections of Street Level Simulations, City Level Simulations, and VRU Simulations, a secondary navigation panel will present options directing users to the SHOW simulation pilot sites. Within these choices, users will encounter further options, each leading to distinct layers of the Simulation Suite, namely Simulating Automated Mobility, Simulation Transferability, and Simulation Levels Connection. This organization essentially mirrors the structure of the simulation suite depicted above, conveying a seamless transition from the tool framework to the web page format.

A number of fruitful aspects stem from the development of the SHOW Simulation Suite for external users, be they researchers or practitioners. Specifically, for each layer, the following generalization and practical capabilities are indicatively provided:

- Users can consult the *Simulating Automated Mobility* layer in order to acquire initial details of the approach of simulating automated mobility, increase their intuition on the topic and become aware of the potential advantages but also caveats of each approach. Moreover, they can become acquainted with some of the horizontal high-level results of CCAM simulations, which can in turn formulate and hone their own research and application scopes more accurately.
- Users can exploit results of the *Simulation Transferability* layer in order to peruse the followed methodologies and to examine more specialised mathematical/ modelling approaches in detail, informing them of the solutions adopted within the SHOW project and thus setting a well-established precedent for CCAM simulation. Furthermore, detailed transferability capacities are showcased, allowing the projection of existing simulation outcomes to other cases, while also providing guidance for shifting from microscopic to macroscopic levels and some tips for combining different available simulation software packages.

- Users can browse the contents of the *Interplay between Simulations* layer provides applied examples for shifting between different scale levels, showing the obtained outcomes to be expected of such processes. This layer also provides comparisons of different tools. These outcomes can be information sources for users wishing to select between available simulation packages.

In addition, the user will receive a multifaceted supply of information- related and stemming from the addressed SHOW pilot sites contexts—that will help them both (i) to navigate and exploit better the contents of the Simulation Suite (e.g. Scope, Key Inputs and Outputs, Strengths and Limitations) and (ii) to learn about the characteristics of the utilized networks and adapt and cross-compare them to the tool users' own networks, in order to create more trustful CCAM Simulation Scenarios. Possible restrictions may relate to the available transport modes in each network, its overall size, and the type of CCAM operation (i.e. mixed traffic environment, dedicated lanes, confined environments).

## 4.3   Simulating Automated Mobility

This layer of the Simulation Suite includes general information on simulating auto- mated driving. Specifically, after selecting the desired simulation category and use case, the simulation suite user is able to be informed about the scope of the studied automated mobility use case, the implemented simulation tool, the key inputs and outputs, strengths and limitations of the followed methodology, the models that were applied and the results of the respective simulation. A step-by-step tutorial layout is incorporated for each scenario and use case, so as to accelerate comprehension by third parties and junior researchers.

- **Scope**: Along with the scope of the studied use case, the most important infor- mation is given to the user regarding the importance of this use case investigation (i.e. relevance to real-world challenges, potential to demonstrate the capabilities and benefits of autonomous vehicles, etc.) the detailed description of the respec- tive pilot site implementation (i.e. network specifications, automated vehicles parametrization, etc.) as well as the selection of the simulated scenarios (i.e. which are the scenarios, why these scenarios were selected for this kind of investigation, etc.).
- **Tools**: Details on the tools employed in this context and a thorough discussion concerning their pertinent technical specifications is provided. This also includes an in-depth discussion of why these particular tools were chosen for the CCAM implementation, highlighting the unique features and capabilities of the tools that contribute to the overall effectiveness of the conducted simulations.
- **Key inputs and outputs**: For simulating automated mobility use cases, various inputs are required. These include real-world traffic data, road network layouts, vehicle specifications, behavioral models, environmental conditions, and user

preferences. For instance, historical traffic flow data from a city, road maps indicating lane configurations, vehicle acceleration and deceleration capabilities, and user demand patterns can all serve as input data for the simulation. Those, as applicable for each use case and scenario, are defined here. Additionally, the simulation generates a range of outputs, including vehicle trajectories, travel times, fuel consumption, emissions, and user satisfaction metrics. Visualization of simulated traffic patterns, statistics on average travel times for different routes, and analysis of carbon dioxide emissions are some examples of the outputs produced by the simulation.

- **Strengths and limitations**: Strengths and limitations of the simulation may vary depending on the specific use case and scenario being analyzed. One strength of the simulation could be its integration of real-world traffic data, which provides realistic simulation outcomes. Incorporating actual traffic flow data allows for an accurate representation of congestion patterns and dynamic traffic conditions. A limitation of the simulation could be the lack of pedestrian traffic modeling, which may underestimate potential conflicts. Ignoring pedestrian behavior could lead to an underestimation of safety hazards and unrealistic simulation outcomes in urban areas with heavy foot traffic.
- **Models**: In the simulation, various models are employed to simulate individual vehicle behavior and decision-making regarding lane changes. Models employed in the simulation may vary depending on both the specific use case and the simulation tool utilized. For instance, the Intelligent Driver Model (IDM) may be used as the car-following model to simulate vehicle acceleration and braking behavior, while the MOBIL model could be employed to simulate lane-changing manoeuvers based on perceived benefits.
- **Results**: Results from the simulation are primarily focused on quantifying the impact of automated mobility services on various metrics related to traffic (e.g., traffic flow, average speed, etc.), environment (e.g., traffic emissions, energy use, etc.) and safety (e.g., traffic conflicts, accidents, etc.). The list of all metrics can be found in SHOW Deliverable 9.2 (Anund et al. 2020).

Focusing on the site interface, as for example, the tab selection interface showcasing the Madrid pilot site navigation is displayed in Fig. 2.

## 4.4 Simulation Transferability

In this layer of the Simulation Suite, which is the second layer as shown in Fig. 1, more technical information is given to the user, i.e. specifications of the followed models. Specifically, the mathematical definitions as well as parametrization and the possibility of transferability is discussed for the models used in the simulation procedure. This will be helpful by giving insights and capabilities regarding traffic simulation in general as well as about automated mobility in specific.

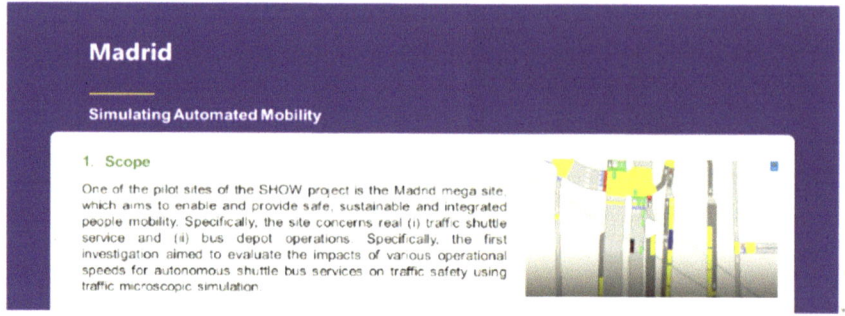

**Fig. 2** Madrid pilot site: scope. *Source* Authors' own picture

- **Mathematical definitions**: In traffic simulations, various mathematical definitions are utilized to model vehicle movement, traffic flow, congestion, and other
  related phenomena. For instance, some key mathematical definitions commonly
  used in traffic simulations are traffic flow models, car-following models (e.g.,
  the Intelligent Driver Model) and lane-changing models (e.g., MOBIL model).
  Hence, the models used in each simulation use case are included in this part along
  with the specific formulas, variables, and guidelines for usage, enabling comprehensive analysis and simulation of various traffic scenarios. As an example, Fig. 3
  shows the mathematical definitions of the applied models within Madrid pilot site
  simulations.
- **Parameterization**: The parameters used for modelling AD vehicles are presented
  as there are many differences between modelling human-driven vehicles and
  AD vehicles and, therefore, by these specifications, an in-depth understanding

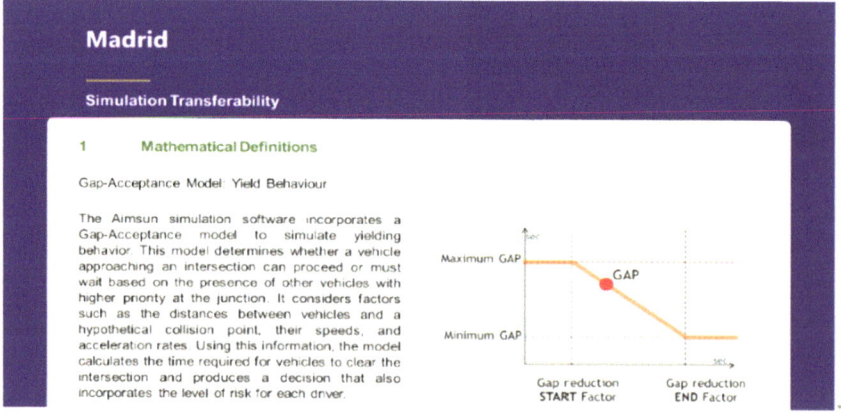

**Fig. 3** Madrid pilot site: mathematical definitions. *Source* Authors' own picture

of modelling AD vehicles will be accomplished (e.g., acceleration and deceleration profiles, reaction time, sensor characteristics, control algorithms, vehicle dynamics, etc.).

- **Transferability**: A discourse on any potential and specific type of transferability of the outcomes across varying simulation scenarios is also included. It is fundamental to investigate and present how the different simulations of each simulated pilot site and their outputs can potentially be combined, as well as how the followed models and indicators could be transferred. As established, traffic flow models can be classified as microscopic, mesoscopic or macroscopic; indeed, within the SHOW project, simulations across all three modelling scales were conducted. The macroscopic models (and mesoscopic models as well) employ aggregated parameters on velocity, density and flow, while microscopic models consider individual vehicle behavior. The vehicle-level simulations are a separate category, as the approach is different from the one that depicts micro–macro (or micro-meso) simulations combination, in order to be combined with the remaining simulation models. The combination of simulations requires an upscaling from microscopic simulations to macroscopic ones as well as from vehicle-level simulations to microscopic simulations in order for a holistic impact assessment of automated fleets to be realized. This upscaling procedure can be realized either through strict mathematical transformations (e.g. Cardaliaguet and Forcadel 2021; Forcadel and Zaydan 2016; Helbing 2007) or by identifying traffic flow parameters or indicators that could be transferrable from microscopic simulations to macroscopic ones using the Macroscopic Fundamental Diagram (MFD). Such indicators include Passenger Car Units—(PCUs, e.g. Tympakianaki et al. 2022), speeds (e.g. Zheng et al. 2017) and headways (Li and Chen 2017).

As an example, Fig. 3 shows the mathematical definitions of the applied models within simulation. As there are differences between modelling human-driven vehicles and AD vehicles there is a need for an extensive understanding of modelling AD vehicles and thus makes this discussion invaluable in comprehending the intricacies and disparities inherent in traffic simulation and consequently is considered as a critical aspect of the Simulation Suite.

## 4.5  *Interplay Between Simulations*

The third and the final layer of the tool, is designed to actually present the followed procedure and results of the application of any interplay between simulations. Furthermore, if any transferability methods proposed in the previous layer (second layer) were utilized, their respective results and steps are also delineated within this layer. Specifically, the given information of this layer serves as a comprehensive guide, aiming to elucidate the process of combining outputs from the different pilot sites simulations, addressing the methodologies employed, and navigating the prospects of transferring these approaches across simulations.

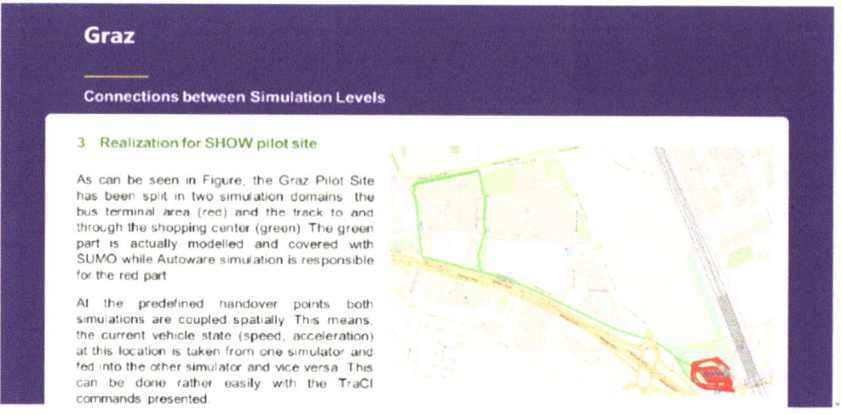

**Fig. 4** Graz pilot site: transition between simulation levels. *Source* Authors' own picture

- **Transition between simulation levels**: The information included will delve into the concept of transition levels, offering insights into the upscaling process, whether from microscopic to macroscopic simulations or vice versa, essential for a comprehensive impact assessment of automated fleets. This involves the exploration of strict mathematical transformations and the identification of transferable traffic flow parameters or indicators utilizing tools.
- **Comparison of different tools**: Additionally, a section comparing different simulation tools used across sites in cases that was applicable provides a comprehensive analysis for combining simulations that utilized varied simulation tools, ensuring a cohesive understanding of the diverse methodologies employed.

For instance, Fig. 4 illustrates the approach taken at the Graz site, where different simulation tools are interconnected to validate the pilot site. In Graz, two simulation levels are utilized to assess safety risks and traffic dynamics. Street level simulations, conducted with SUMO, focus on traffic flow and congestion without considering vehicle sensors or environmental occlusions. Meanwhile, VRU-level simulations, employing AWSIM, simulate vehicle sensors realistically within a 3D environment, reducing the gap between simulation and reality. Integrating these simulation levels offers a comprehensive understanding of traffic dynamics while minimizing discrepancies between simulation and real-world scenarios. This chapter discusses the process of coupling SUMO with other simulators to leverage their combined advantages effectively.

## 4.6  Library of Simulations

The SHOW simulation Library is created to be the static repository of fundamental information regarding simulating automated mobility of each Simulation Suite layer.

More specifically, it includes all important data in appropriate format in order to be easily downloaded and used by the user in their tutorials/educational exercises. This kind of data is:

- **Raw simulation data**: Raw results extracted from the traffic simulation tool, which enable further filtering and processing in an alternative manner.
- **Scripts/APIs**: Includes scripts and/or APIs used in the respective simulation, which allow users to directly utilize relevant scripts for similar cases or within the same simulation tool.
- **Simulation visualization**: Recorded videos and screenshots from the simulation procedure that aid in familiarizing users with the simulation software.
- **Visualized guidelines**: Offers visualized guidelines across different use cases and software.
- **Documentation**: Relevant documentation for automated mobility use cases including research papers, traffic simulation instructions or tutorials, theoretical background documentation of behavioral models and algorithms, etc.

## 5 Conclusions

The core objective of the SHOW simulation suite is to seamlessly merge the insights garnered from simulating automated mobility, ensuring optimal integration across various simulation levels. This encompasses scalability, robust data analysis, and the cohesive synthesis of fundamental aspects to achieve a comprehensive understanding. This is accomplished by the development of the web-tool that is presented in this chapter, which is also considered to form the simulation suite scope.

With regards to the added value on the project level, the simulation suite tool will lead to the exploitation and dissemination of SHOW findings at the maximum possible degree. Furthermore, significant advancement will be made as in the tool all possible data and information are collected, enhancing in this way the data availability for each site. Specifically, this process gives deeper insights into the pilot sites with indicators that cannot be directly measured in real-life pilot sites as well as make comparisons between real and simulation data. In addition, by manipulating critical aspects of the simulation in a similar manner for all pilot sites, comparisons among scenarios, networks, models, methodologies and tools could be easier generated. Last but not least, with the proposed up-scaling capabilities, there are many benefits for the project. One fundamental advantage lies in the integration of diverse simulation levels (microscopic and macroscopic), yielding generalized outcomes rather than independent ones, each resulted from distinct objectives. Another fundamental benefit is that by up-scaling data, in many cases there will be the possibility to generate more detailed data for automated mobility. This means that as the volume of data increases, it allows for the creation of more detailed and nuanced insights into automated mobility systems and their performance. For instance, by collecting data from a larger geographic area or a greater number of vehicles, researchers can better understand the intricacies of AVs behaviors and interactions, leading to more informed decision-making and improved system design.

From the user side and beyond the project, the SHOW Simulation Suite is useful for every researcher who is interested in simulating automated mobility, being and expert or not in traffic simulation. For this reason, the suite is designed in order to provide information about the possible tools and layers, suitable scenarios, and guidelines and to further give directions about the user's desired simulation scenario or use case or study area. Moreover, as already mentioned in the previous sections, more mathematical information are given by using the tool as well. This means that simulation experts will also gain knowledge through the documentation provided by the tool, as automated mobility is under investigation and hence there are significant challenges in simulating automated driving.

Finally, the included SHOW simulation library could make the tool also beneficial for city planners as well as practitioners, such as professionals actively involved in urban planning, policy implementation, or related fields who can utilize the simulation suite to inform their decision-making processes and operational strategies. Key results from each analyzed use case and level are archived, enabling interested stakeholders to guide future city management through appropriate strategies. Therefore, the simulation suite could also guide interested stakeholders in the future management of cities by using suitable strategies, as transportation systems will be fundamentally affected by the evolution of automated driving.

**Acknowledgements** The research is part of the SHOW project that has received funding from European Union's Horizon 2020 research and innovation programme under grant agreement no. 875530.

# References

Anund A et al (2020) SHOW Deliverable 9.2: Pilot experimental plans, KPIs definition & impact assessment framework for pre-demo evaluation. https://show-project.eu/wp-content/uploads/2022/10/D9.2-Pilot-experimental-plans-KPIs-definition-impact-assessment-framework-for-pre-demo.pdf

ARCADE (Aligning Research & Innovation for Connected and Automated Driving in Europe) project. Funding from the European Union's Horizon 2020 research and innovation programme. https://www.connectedautomateddriving.eu

Cardaliaguet P, Forcadel N (2021) From heterogeneous microscopic traffic flow models to macroscopic models. SIAM J Math Anal 53(1):309–322. https://doi.org/10.1137/20M1314410

European Road Transport Research Advisory Council (ERTRAC) (2019) Connected Automated Driving Roadmap. https://www.ertrac.org/wp-content/uploads/2022/07/ERTRAC-CAD-Roadmap-2019.pdf

Forcadel N, Zaydan M (2016) Derivation of a macroscopic LWR model from a microscopic follow-the-leader model by homogenization. IFIP Adv Inf Commun Technol 494:272–281. https://doi.org/10.1007/978-3-319-55795-3_25

Helbing D (2007) From microscopic to macroscopic traffic models. Perspect Look Nonlinear Media 122–139:2008. https://doi.org/10.1007/bfb0104959

LEVITATE (Societal Level Impacts of Connected and Automated Vehicles) project. Funding from the European Union's Horizon 2020 research and innovation programme. https://levitate-projec t.eu/

LEVITATE connected and automated transport systems PST (Policy Support Tool). https://www. ccam-impacts.eu/

Li L, Chen X (Michael) (2017) Vehicle headway modeling and its inferences in macroscopic/ microscopic traffic flow theory: a survey. Transp Res Part C: Emerg Technol 76:170–188. https:// doi.org/10.1016/j.trc.2017.01.007

Martensen H, Diependaele K, Daniels S, Van den Berghe W, Papadimitriou E, Yannis G, Elvik R (2019) The European road safety decision support system on risks and measures. Accid Anal Prev 125:344–351. https://doi.org/10.1016/j.aap.2018.08.005

SafetyCube (Safety CaUsation, Benefits and Efficiency) Funding from the European Union's Horizon 2020 research and innovation programme. https://www.safetycube-project.eu/

SafetyCube Decision Support System. https://www.roadsafety-dss.eu/

SHOW (SHared automation Operating models for Worldwide adoption) project. Funding from the European Union's Horizon 2020 research and innovation programme, GA No 875530. https:// show-project.eu/

Ziakopoulos A, Roussou J, Chaudhry A, Hu B, Zach M, Oikonomou M, Veisten K, Hartveit KJ, Brackstone M, Vlahogianni E, Thomas P, Yannis G (2022) The LEVITATE policy support tool of connected and automated transport systems. In: Proceedings of the 9th transport research arena TRA 2022. Lisbon, Portugal. https://doi.org/10.1016/j.trpro.2023.11.566

Tympakianaki A, Nogues L, Casas J, Brackstone M, Oikonomou MG, Vlahogianni EI et al (2022) Autonomous vehicles in urban networks: a simulation-based assessment. Transp Res Record 03611981221090507. https://doi.org/10.1177/03611981221090507

Zheng L, He Z, He T (2017) A flexible traffic stream model and its three representations of traffic flow. Transp Res Part C Emerg Technol 75:136–167. https://doi.org/10.1016/j.trc.2016.12.006

# On International Collaboration Within Research Projects for Large-Scale Piloting of Shared Automated Mobility

**Henriette Cornet, William Riggs, Maria Gkemou, and Stephane Dreher**

**Abstract** Automated vehicles (AVs) are being tested and deployed globally, offering benefits such as reduced accidents, enhanced public transportation, and optimized operational costs. These advancements vary by region, with Europe focusing on shared AV fleets in public transport, Japan investing in infrastructure and safety technologies, and the U.S. being mostly driven by private sector innovations. This paper highlights the importance of international collaboration in AV development and deployment, emphasizing the need for thorough planning before any mutual project initiation. Effective collaboration so far is revealed to be more feasible and fruitful at the research level rather than high policy levels, especially when projects have similar scales, scopes, and timelines. Despite resource constraints, international collaboration is crucial for exchanging on practices and will become more and more important for harmonizing regulations and bringing innovation around AV technology. This chapter explores best practices and lessons learned for collaboration between the EU, the U.S., and Japan, using case studies built in SHOW (GA No 875530), to guide future international efforts in sustainable automated transport.

**Keywords** International collaboration · Research projects · Funded projects · Large-scale pilots · Automated mobility

H. Cornet (✉) · W. Riggs
University of San Francisco, 2130 Fulton St., San Francisco, CA 94117, USA
e-mail: hcornet@usfca.edu

W. Riggs
e-mail: wriggs@usfca.edu

M. Gkemou
Centre for Research & Technology Hellas (CERTH), Hellenic Institute of Transport (HIT), 34 Ethnarchou Makariou St., 16341 Athens, Greece
e-mail: mgemou@certh.gr

S. Dreher
ERTICO, Av. Louise 523, 1050 Brussels, Belgium
e-mail: s.dreher@mail.ertico.com

© The Author(s) 2025
H. Cornet and M. Gkemou (eds.), *Shared Mobility Revolution*, Lecture Notes in Mobility, https://doi.org/10.1007/978-3-031-71793-2_14

233

# 1 Introduction

In numerous countries over the world, automated vehicles (AVs) are being tested and deployed as a new paradigm for transport. The technology offers the potential to reduce accidents caused by human error, to enhance public transportation systems with a higher quality of service (flexible and on-demand) and to optimize the cost of operations. These steps toward shared automation are being achieved in different ways around the globe and they lead to opportunities for international knowledge sharing, collaboration and partnership.

In Europe, with large projects like AVENUE,[1] SHOW,[2] and ULTIMO,[3] many deployments are anchored within the public transport context, having fleets of shared vehicles mostly along predefined urban routes, building upon past C-ITS program investments. By contrast, AV deployments are different in Japan and the U.S. In Japan, AV development programs have focused substantially on infrastructure investments and on increasing safety and technological advancement in Advanced Driver Assistance Systems (ADAS) technology and connectivity as well as on improving mobility to address societal challenges in rural areas. In the U.S., AV technical development has been funded by the companies that are developing the products. Product development has included commercial vehicles, local delivery vehicles, transit-oriented shuttles, and ride-hailing services with companies like Cruise and Waymo fielding robotaxis in places like greater Phoenix and San Francisco. There has also been a focus on specific use-cases for long-haul trucking, goods delivery and shared AVs for those with disabilities, in the form of paratransit and wheelchair accessible vehicles (Federal Transit Administration 2022).

In this context, international collaboration and knowledge sharing are essential to reap the benefits of automation while mitigating potential negative impacts and externalities. Indeed, collaboration on AVs can help harmonize regulations, standards, and practices in a way that accommodates the complexity of automated systems but maintains a grounding in citizen dialogues and stakeholder needs (Chng et al. 2021).

Global knowledge sharing can help streamline development of AVs, including experiences and technological breakthroughs, and business strategies. It can enable regions to learn from one another's pilot projects and policy experiments, such as Europe's focus on public transport-centric AV deployment or the United States' emphasis on demand-responsive rideshare platforms led by the private industry.

Given the benefits of collaborative approaches, the aim of this chapter is to explore best practices for collaboration and exchange on the global level with a focus on collaboration between the European Union, the United States and Japan.

This context is used to describe specific collaborations across these geographies. Each collaboration is described and then summarized with core takeaways bulleted so that they can be easily transferred. The goal of this is to facilitate and provide

---

[1] https://www.h2020-avenue.eu/.

[2] https://www.show-project.eu/.

[3] https://www.ultimo-he.eu/.

lessons for future international collaborations to facilitate the future of sustainable automated transport.

## 2 Focus of Pilot Research in EU, U.S. And Japan

### 2.1 Focus of Pilot Research Projects in Europe

The largest EU projects on automation receive funding from either the Horizon 2020 programme (2014–2020, €80 billion funding for the entire programme) and, for 2021–2027, the Horizon Europe programme with €95.5 billion funding (European Commission 2020). Calls for proposals are released regularly by the European Commission (EC) and organizations can form consortia and submit project proposals in response to these calls. The EC evaluates proposals based on criteria like innovation and impact, selecting the most promising ones for funding.

For instance, SHOW (Horizon 2020, GA No 875530) answered in 2019 the call "Developing and testing shared, connected and cooperative automated vehicle fleets in urban areas for the mobility of all" (DT-ART-04-2019) aiming at deploying 70 AVs for passengers and goods in 20 cities across Europe. Following the AVENUE project funded by Horizon 2020, SHOW was followed by ULTIMO under Horizon Europe, all answering in similar EC topics.

These projects focus on shared automated vehicles (AVs) in urban and peri-urban environments, are distinct from endeavours like L3-Pilot[4] or Hi-Drive,[5] which focus on private vehicles operating in a variety of settings including urban roads and highways. There are also complimentary projects that are non-operational level, such as the FAME project,[6] which aims to consolidate knowledge within the CAD Knowledge Base, an online repository that gathers the information from among a broad network of CAD stakeholders to establish a common baseline of knowledge related to R&I, testing and demonstration of CAVs in Europe (CAD Knowledge Base 2024).

At their core, EU projects aim to strengthen European industry and competitiveness by creating collaboration within consortia of stakeholders from industry, academia and associations or public entities based in Europe. This unique setting with funding allocated by the EC—in most cases however, anticipating a co-funding capability by the participating industrial actors—allows each stakeholder to pursue their own business goals through robust Consortium Agreements. It also ensures the involvement of a variety of EU countries and aims at facilitating knowledge transfer between the private and public sectors.

An important common feature of these EU large-scale pilot projects dedicated to urban mobility is the use of shuttles. European OEMs were at the forefront as

---

[4] https://www.l3pilot.eu.

[5] https://www.hi-drive.eu/.

[6] https://www.connectedautomateddriving.eu/about/fame/.

AVENUE and SHOW started, with deployments from Navya (now Gama), EasyMile and Sensible4. Indeed, in SHOW, from the 80 vehicles deployed for passengers and cargo, 10 were from Navya/Gama, 18 from EasyMile, 2 from Sensible4, the rest being vehicles retrofitted by industry or research institutes.

Apart from pilot projects funded by the EC, other pilot or pre-commercial initiatives on AVs are occurring on national or regional level in Europe, in which cases funding may also emerge from national or governmental public funds or by respective Public Private Partnerships formed for this scope.

While the European Commission prioritizes allocating funding within European borders, there is a clear impetus for collaboration with international entities. This is exemplified by the partnerships forged by SHOW with counterparts in the United States and Japan.

## 2.2   Pilot Research Projects in the U.S.

According to the Alliance for Automotive Innovation, as of 2022 there were 84 AV companies operating across 30 states and 120 cities in the U.S.—many of which were working across industry sectors with international footprints (Alliance for Automotive Innovation 2022).

There are no large scale publicly funded demonstrations of automated vehicles in the U.S. that are comparable to those funded by the European Commission. Most automated vehicles operations in the U.S. are funded by the private AV companies as part of their product development. However, there are many small projects in the U.S. developed by universities and transport agencies with support from the federal U.S. Department of Transportation (USDOT), the National Highway Traffic Safety Administration (NHTSA) or the Federal Transit Administration (FTA). This government-funded research has concentrated primarily on local initiatives and has focussed on specific topics such as accessibility for automated public transit.

For example, government funding has been provided for the pilot of an automated bus from ADASTEC on the Michigan State University in Michigan (Michigan State University 2021), to ADS for Rural America from the University of Iowa (The University of Iowa 2024) and to the Contra Costa Transportation Authority (CCTA) for its automated shuttle pilots in California (Contra Costa Transportation Authority 2024). Also, AV pilots like the Arlington RAPID program received a $1.7 million grant through the Federal Transit Administration's Integrated Mobility Innovation Program to integrate a fleet of five autonomous vehicles into its Via on-demand rideshare service (City of Arlington 2024). Similarly, the goMARTI project in Grand Rapids, Minnesota offers free on-demand rides with a fleet of five May Mobility autonomous vehicles, including three wheelchair-accessible vehicles (goMARTI 2024).

In parallel, innovations have emerged through private investment and venture capital, far surpassing the capacity of government funding. According to a 2019 report by McKinsey, $13.5 billion had been invested in AV software and mapping, with an additional $29.9 billion allocated to AV sensors and ADAS components

(McKinsey & Company 2019). Waymo stands out as the primary recipient of funding, with a total of \$5.5 billion, a portion of which comes from Alphabet, i.e., Google (CNBC 2022).

As a result of this investment scale, companies like Cruise, Waymo, Zoox, and others have developed, tested and deployed autonomous fleets across multiple geographies. For instance, in California, both Cruise and Waymo received all required permits for both driving and passenger carriage activities, and successfully launched a fully self-driving ride-hailing commercial service to clients in 2022 in the city of San Francisco (California Public Utilities Commission).

The fact that AV companies have been able to pilot their vehicle technology and services on public roads, and in real-world conditions, has helped build rich data sets and critical real-world experiences of automated driving. It has also led to public dialogue about some of the challenges with the technology and better ways to address them, with public advice letters provided to agencies detailing varied opinions between many stakeholders on the issue. In California, issues related to in-lane pick-up and drop-off, vehicle recovery due to equipment malfunction, communications redundancy, interactions with law enforcement, incidents response and other corner-case scenarios of AV services have prompted government organizations to hold hearings and workshops to foster dialogues between industries, agencies and the public. For example, the California Public Utilities Commission hosted a public workshop on autonomous vehicle data reporting conducted in June 2023 (Workshop on Autonomous Vehicle Data Reporting).

Independently from government funding, collaborations between these industry players and universities are taking place, contributing to knowledge building and sharing beyond the confines of the private sector. Notably, the "Research Rider" program from the University of San Francisco's (USF) University's Autonomous Vehicles and the City Initiative in 2022. This programme, which was supported by a Memorandum of Understanding (MoU) with Cruise, reported the rides of 300 students over a series of months to understand how it impacted their travel behaviour and to gather rider insights about their perception towards the AV technology and services (Riggs et al. 2023).

All these endeavours show that U.S. industry remains far ahead with commercial deployments by private players such as Cruise and Waymo with extended Operational Design Domain (regarding speed, route types, situation awareness, miles travelled). In this context, fostering transatlantic collaboration becomes crucial and it is imperative to ensure that European efforts remain relevant and not swiftly outdated.

## 2.3  Pilot Research Projects in Japan

As early as 2015, in the Fifth Science and Technology Basic Plan, the Japanese government presented how science and technology innovation policy represents a major policy goal for the Japanese economy, society, and the public (Japanese Cabinet Office). In this plan, a vision for a "boldly challenging the future" was laid out with the concept of becoming a world-leading "super smart society" (so called, Society 5.0), in which Information and Communication Technology (ICT) is being leveraged to its fullest in the manufacturing sector. Core systems of the Society 5.0 include Intelligent Transportation Systems, with a desire to conduct collaborations between industry, academia, and government and a goal to accelerate coordination of multiple systems and improve industrial competitiveness.

In its Sixth Science, Technology, and Innovation Basic Plan from 2021, autonomous driving is explicitly mentioned within the Phase 2 of the Research and Development Themes of the Cross-ministerial Strategic Innovation Promotion (SIP) Program (Japanese Cabinet Office). As a subprogram, the SIP's Automated Driving for Universal Services (SIP-adus) has been progressing on research and development activities aiming to "solve issues of concern in today's society, including reducing traffic accidents, alleviating traffic congestion and securing a means of transportation for people with limited mobility, such as the elderly living in remote regions, among other issues" (SIP-ADUS). Since 2023, the Mobility Innovation Alliance has taken over academic collaboration activities of the SIP-adus explicitly promoting collaboration activities with overseas research institutes and projects (Mobility Innovation Alliance). Among others, the Mobility Innovation Week Japan event aims to gather international experts on a yearly basis on the theme of innovation in mobility including autonomous driving.

Regarding autonomous driving demonstrations on public roads, in 2021, the Japanese Ministry of Economy, Trade and Industry (METI) and Ministry of Land, Infrastructure, Transport and Tourism (MLIT) jointly launched the "Project on Research, Development, Demonstration and Deployment (RDD&D) of Automated Driving toward Level 4 and its Enhanced Mobility Services", known as "RoAD to the L4" to develop advanced mobility services. Under the RoAD to the L4 project, several initiatives have been planned and executed in different locations (U.S. Department of Commerce).

For instance, in Sakai, BOLDLY Inc., an automated driving systems development and service subsidiary of SoftBank Corp., is operating several Navya shuttles as first and last mile services with local financing of ¥520 million ($5 million) for five years that will end in March 2025.[7] Another noteworthy initiative is the Automated Bus Pilot Deployment at the University of Tokyo's Kashiwa Campus, which commenced in November 2019, covering a 2.6 km route on public roads from the railway station to the university campus. This pilot is part of the project CooL4, focused on Cooperative Level 4 Automated Mobility Service in mixed traffic Environment. CooL4 is led by the University of Tokyo, collaborating with research institutes (e.g., the Mitsubishi research institute) as well as industry (e.g., Panasonic, BOLDLY). Dedicated working teams of CooL4 are investigating international collaboration activities to facilitate deployment of Cooperative Automated Driving Systems—which will be developed in further sections of this chapter.

---

[7] https://www.softbank.jp/en/sbnews/entry/20210622_01.

# 3 International Collaboration Efforts and Best Practices

## 3.1 Background of International Collaboration in Research Projects

The European Commission (EC), the United States Department of Transportation (USDOT), and the Road Bureau of Ministry of Land, Infrastructure, Transport and Tourism (MLIT) of Japan have had a formal collaboration in the area of road automation since 2012. The so-called Trilateral Working Group on Automation in Road Transportation (ART) was part of a broader Trilateral exchange framework aimed at exchanging research and collaborating in high-priority areas related to ITS (Intelligent Transportation Systems).

The ART working group focused on the safety, mobility, and environmental impacts of highly automated vehicles in anticipation of rapid AV development and public roadway operations, and had the following objectives:

- Allow each region/country to learn from one another's programs
- Identify areas of cooperation where each region will benefit from coordinated research activities, and
- Engage in cooperative research and harmonization activities, as possible.

The FAME Project and its preceding project ARCADE[8] had among their objectives to support the European Commission in identifying EU-funded projects with strong potential for cooperation with similar initiatives in US and Japan. They organised the contribution of European experts and R&I projects to the ART working group in five areas that were selected as priorities for cooperation: human factors, impact assessment, roadworthiness testing, physical and digital infrastructure (PDI) and next generation transport. While the exchanges addressing the three former topics led to successful project twining activities, common research papers or the development of common methodologies (e.g., Trilateral Impact Assessment Framework for ART[9]), the cooperation in the area of next generation transport, which included demonstrations of shared automated transport, remained limited to an information exchange among the three regions. This was mainly due to the different objectives, drivers, sizes and timelines of shared AV development and demonstration programs, as can be inferred from the regional overviews above. Nevertheless, the trilateral cooperation in this domain led to fruitful discussions in particular on common challenges, such as the identification of win-wins for the next steps, the nature of supervision that will be required for shared AVs, user friendly design and acceptance, or on how AVs interact with the PDI.

---

[8] https://www.connectedautomateddriving.eu/about/arcade/.

[9] https://www.connectedautomateddriving.eu/wp-content/uploads/2018/03/Trilateral_IA_Fram
ework_April2018.pdf.

With this background, the EC encouraged future consortia for international collaboration with requirements that mentioned explicitly in its Horizon 2020 call (below as an example the call DT-ART-04–2019 to which SHOW answered):

> In line with the Union's strategy for international cooperation in research and innovation, international cooperation is encouraged. In particular, proposals should consider cooperation with projects or partners from the US, Japan, South Korea, Singapore, and/or Australia. Proposals should foresee twinning with entities participating in projects funded by US DOT to exchange knowledge and experience and exploit synergies. Twinning with Japan is also encouraged.

In the recent update of its Strategic Research and Innovation Agenda (SRIA[10]), the European Partnership of Connected, Cooperative and Automated Mobility (CCAM) puts a strong emphasis on the importance of international cooperation, in particular to share best practices, address common challenges more efficiently or develop harmonised solutions.

Because of the trilateral discussions, an emphasis was placed from the EC perspective on U.S. and Japan activities and this is how projects like SHOW (and now ULTIMO) have initiated collaborative case studies explicitly with American and Japanese entities. A selection of these case studies is presented hereinafter.

## 3.2 International Collaborations in SHOW with U.S.

Building on the Research Rider program, USF entered into a Non-Disclosure Agreement (NDA) with SHOW to actively share knowledge with partners about the findings related to L4 rideshare activities occurring in San Francisco. While the NDA between USF and SHOW took over a year to be completed, the partnership provided the chance to leverage these insights internationally and to participate in dialogues with European counterparts.

A key activity was a student workshop during which USF students developed stakeholder personas for automated mobility to be used within SHOW Ideathons (Cornet and Riggs 2023). The agenda for the workshop was structured such that after an introduction on the ideas behind automated vehicles, students would engage in ideating a picture of what a stakeholder might look like—aka a "persona". This concept of personas was based on research completed by TUMCREATE on personas and included the following general components: Name, Age, Key Details; Goal / Bio; Quote; Motivations; Frustrations (Kong et al. 2018).

Based on their knowledge from U.S. deployments with companies like Cruise and Waymo and their introduction to the SHOW project, the students were challenged to provide an embodiment of the people who may have a stake in large scale AV deployments—including the perspectives of both advocates/proponents of the technology along with characterization of individual who may have critical needs, not feel comfortable week or even oppose AVs in their community. An example of

---

[10] https://www.ccam.eu/wp-content/uploads/2023/11/CCAM-SRIA-Update-2023.pdf.

**Fig. 1** USF-SHOW persona examples, left: Bob Miller, baker, who needs support for his bread delivery; right: Michelle Brown, school director who needs safer and more sustainable transit options for the school students. *Source* Midjourney's pictures rendered by Cornet and Riggs (2023)

their depictions of a small business owner and a school director from the exercise are provided in Fig. 1.

Following on to this activity, the students were challenged to explore business models and strategies that might meet the need of their stakeholder. In this way the USF-SHOW workshop offered the opportunity to explore ways to reimagine the transport value chain (vehicle production, automated car technology, fleet management and operations, and B2C mobility services), and look at potential business opportunities and models in this space.

The USF-SHOW workshop exemplifies effective collaboration between U.S. universities and an EU-funded project. It required minimal organizational resources while fostering bilateral learning. USF gained insights into the strong orientation of autonomous vehicles (AV) for public transport in EU, while SHOW benefited from a fresh perspective on how AV services tested in the project can impact different demographics.

In addition to USF, SHOW engaged with U.S. counterparts through a webinar with the USDOT Volpe Center[11] to discuss the accessibility of AV services. The Volpe Center brought together experts from the Santa Clara Valley Transportation Authority (California), Contra Costa Transportation Authority (California), and the City of Arlington (Texas). SHOW presented pilot sites focused on AV service accessibility in Frankfurt am Main (Germany), Tampere (Finland), and Carinthia (Austria). While the exchange was fruitful, no formal follow-up or documentation of results occurred.

Other collaborations took place during international events where the SHOW consortium invited experts from U.S. to join discussions during panels, e.g., at the TRB annual Automated Road Transportation Symposium[12] in U.S., EUCAD[13] in Belgium and the Mobility Innovation Week[14] in Japan.

---

[11] https://www.volpe.dot.gov/.

[12] https://www.trb.secure-platform.com/a/page/AutomatedRoadTransportationSymposium.

[13] https://www.cinea.ec.europa.eu/news-events/events/eucad2023-4th-european-conference-connected-and-automated-driving-2023-05-03_en.

[14] https://www.mobilityinnovationalliance.org/mobiweekjapan2023/en/.

## 3.3   International Collaborations in SHOW with Japan

Japan engaged in collaborating with projects funded by the European Commission, with partnerships established with SHOW, ULTIMO, and SUNRISE[15] for addressing, among others, AV driving behaviour, regulatory aspects, and stakeholders' perceptions of shared AV services through webinars and joint conference sessions.

Within SHOW, the collaboration with CooL4 was finalized in October 2022 with the signature of an MoU, and the following topics were identified for collaboration:

- **Topic 1: Automated Vehicle Behaviour**: Addressing challenges at intersections during GPS-loss, this topic explores strategies to navigate around parked vehicles. Additionally, it investigates the integration of Vehicle-to-Everything (V2X) communication and the implementation of remote supervision and control mechanisms.
- **Topic 2: Regulatory Issues**: Focusing on the legal aspects of SAE Level 4 vehicles, this topic delves into the legal framework, type approval, and homologation processes. The ultimate goal is to propose modifications to existing regulations and define international rules to address liability concerns.
- **Topic 3: Passengers' Perception of AV Services**: The goal of this topic is to conduct a cross-country comparison of stakeholders' perceptions of automated vehicle services. It also aims at exploring various business models and services within the automated vehicle sector.

After agreeing on the topics and identifying expert teams willing to allocate resources for the collaboration, a series of webinars was organized, gathering the respective teams of experts for each topic. The collaboration operated under a resource-efficient approach, where each party contributed their own resources. Topic 1 was organized with 23 experts from EU and 20 experts from Japan as an online webinar. The webinar covered a range of topics from both the EU and Japan, focusing on advancements and applications in autonomous driving technology. The EU presented on using sensors to enhance decision-making, the third-generation advancements of EasyMile's EZ10 people mover in perception, localization, and navigation, and the challenges of implementing autonomous driving in mixed traffic environments. Japan's contributions included an introduction to CooL4 activities related to technology, the Cooperative Traffic Safety System (CTSS), high-speed image processing, sensor network systems, reliable communications for cooperative systems, ITS communication, localization, and measures in road space. Following the webinar, a Q&A (questions and answers) document was exchanged to complete the understandings on traffic signal information and infrastructure sensing devices. Due to the confidentiality of the information discussed during this webinar, the documentation cannot be shared publicly.

For topic 2, the event's agenda covered a range of topics, including the EU Automated Driving System (ADS) regulation (presented directly by a member of the EC)

---

[15] https://sunrise-europe.eu/.

and C-ITS service validation as a basis for AV testing and certification schemes. Discussions also addressed local policy interventions for deploying shared automated vehicles in urban areas, insights from the CooL4 project, and Japanese laws concerning automobiles and the legislative process for automated driving. Additionally, the agenda included analysis of the 2019 and 2022 amendments to Japanese laws, future issues, and challenges related to cooperative automated driving. Topic 3 was primarily addressed through informal discussions aimed at defining the scope of the study. However, no concrete outcomes were reached by the time this chapter was written.

Similarly to the US, the collaboration also took form within the organisation of joint sessions at conferences and global events such as the TRB annual Automated Road Transportation Symposium in U.S., EUCAD in Belgium and the Mobility Innovation Week in Japan.

Furthermore, Japanese delegations visited pilot sites in Europe, among others from SHOW in Linköping (Sweden), Monheim (Germany) and Brno (Czech Republic).

## 3.4  Lessons Learned and Recommendations from SHOW

The collaborations between SHOW and U.S. and Japan entities have led to lessons that have applicability for future projects, ranging from activating individual stakeholders to decisions that can be made when contemplating project designs. Table 1 summarizes primary recommendations for collaboration that can be beneficial for future research projects and pilot AV deployments—specifically those aiming at large-scale demonstrations.

In addition to these lessons learned from the SHOW project, the international exchanges in ARCADE, FAME and the Trilateral EU-US-Japan cooperation activities have brought forward the need to maintain exchanges both at institutional level and expert level for an efficient cooperation. A regular exchange among funding institutions from the different regions is essential to enable an early alignment on possible R&I roadmap topics for cooperation. Funding programmes can ideally include the requirement for specific future projects to cooperate with similar initiatives in another region, enabling the projects to plan ahead and optimise milestones in the initial phase. The continued exchange of knowledge among experts and presentations of project plans even before the start of actual developments in the frame of international conferences like EUCAD, ARTS and the Mobility Innovation Weeks is fundamental for the early identification of common harmonisation needs or solutions to common challenges, thus optimising funding and resources.

**Table 1** Core Takeaways from EU, US and Japan Collaborations in SHOW

| Takeaway | Description |
| --- | --- |
| Loose structures | Less formal consortia with flexible memberships create greater inclusivity and more opportunity for new ideas to emerge |
| Reconsider the need for formal agreements | Formalizing an MoU can be time-consuming, and often, a simple NDA is sufficient to facilitate meaningful discussions among researchers |
| Specificity of the topics and alignment in time | The field of investigation for international collaboration should be narrow enough to enable deep discussions between researchers rather than high-level exchanges. Furthermore, collaboration is easier when projects are at the same stage of development, such as during the exploration phase or when the vehicles are on the road |
| Individual relationships at the project level are key | While trilateral agreements may be effective at policy levels, matching individuals to share highly granular aspects of technology, operations, businesses models and more is key to technical advancement and knowledge sharing |
| Funding schemes allowing for structured international collaboration—also for small projects | Specific and structured funding schemes, along with mutually agreed goals and expected collaboration items, are seen as clear enablers from and for all parties. This planning should occur before the project starts, not as an afterthought. Not all projects need to be large-scale; smaller, focused projects can be developed and deployed more quickly to address very specific questions |
| Standardization in research | Standardizing research methodologies and developing interoperable data formats is important to enable the merging of data sets, which can lead to larger, comparable results and potentially the creation of "twins" between tests, for instance, using the same impact assessment framework, and thereby enhancing the robustness of findings and recommendations |

# 4 Conclusion

In advancing automated mobility, particularly within the realm of public transport, international collaboration is paramount for creating synergies and facilitating knowledge and innovation transfer across continents. Despite varying investment strategies and scopes defined among countries, shared learnings from different AV pilots can provide valuable insights, preventing the need to begin anew and to duplicate efforts among industry, academia, and government sectors or to lose global momentum.

While establishing collaboration beyond a project's scope poses challenges— often due to underestimated initial efforts, asynchronous project timelines and different scales—the value of sustained, durable relationships cannot be overstated. Knowledge can be built and shared incrementally through regular interactions, benefiting all participants. The chapter has exemplified this fact with case studies from the international collaboration in SHOW with U.S. and Japanese entities, happening mainly at academic and research level but not only.

Looking ahead, to support global advancements and market integration, there will eventually need to be harmonization between regulatory frameworks and clear insight on comparable business models and international collaboration can help for that. Companies and public entities from regions like the EU, U.S., and Japan seek to operate globally and should not be expected to constantly adapt to varying communication protocols, data definitions, and deployment regulations. Building on research collaborations, the establishment of unified regulations will be essential for the seamless, interoperable and efficient deployment of shared automated mobility solutions worldwide that benefit all of the diverse stakeholders and consumers that our public transportation systems serve.

# References

Alliance for Automotive Innovation (2022) Ready to launch autonomous vehicles in the U.S. Tracking the current (and future) AV landscape. Alliance for Automotive Innovation, Washington D.C. [Online]. Available: https://www.autosinnovate.org/posts/papers-reports/AV%20Report.pdf

CAD Knowledge Base (2024) About [Online] Available: https://www.connectedautomateddriving.eu/about/

California Public Utilities Commission, Autonomous Vehicle Program Permits Issued [Online]. Available: https://www.cpuc.ca.gov/regulatory-services/licensing/transportation-licensing-and-analysis-branch/autonomous-vehicle-programs/autonomous-vehicle-program-permits-issued

City of Arlington (2024) Texas, Autonomous Vehicles, consulted 18 May 2024 [Online]. Available: https://www.arlingtontx.gov/visitors/transportation/autonomous_vehicles

Chng S, Kong P, Lim PY, Cornet H, Cheah L (2021) Engaging citizens in driverless mobility: Insights from a global dialogue for research, design and policy. Transp Res Interdisc Perspect 11:100443

CNBC (2022) Why the first autonomous vehicles winners won't be in your driveway [Online]. Available: https://www.cnbc.com/2022/05/21/why-the-first-autonomous-vehicles-winners-wont-be-in-your-driveway.html

Contra Costa Transportation Authority (2024) Automated Driving Systems, consulted 18 May 2024 [Online]. Available: https://ccta.net/projects/innovate-680/automated-driving-systems/

Cornet H, Riggs W (2023) Personas for autonomous vehicle deployments. Rochester, NY. https://doi.org/10.2139/ssrn.4491033

European Commission, Horizon 2020, consulted on May 1, 2024. [Online]. Available: https://research-and-innovation.ec.europa.eu/funding/funding-opportunities/funding-programmes-and-open-calls

Federal Transit Administration (2022) Accessibility in transit bus automation: scan of current practices and ongoing research, FTA Report No. 0228 [Online]. Available: https://www.transit.dot.gov/about/research-innovation

goMARTI (2024) Experience self-driving in Minnesota's nature! consulted 18 May 2024 [Online]. Available: https://www.gomarti.com/

Japanese Cabinet Office, Report on The 5th Science and Technology Basic Plan [Online]. Available: https://www8.cao.go.jp/cstp/kihonkeikaku/5basicplan_en.pdf

Japanese Cabinet Office, Science, technology and innovation basic plan [Online]. Available: https://www8.cao.go.jp/cstp/english/sti_basic_plan.pdf

Kong P, Cornet H, Frenkler F (2018) Personas and emotional design for public service robots: A case study with autonomous vehicles in public transportation. In: 2018 international conference on cyberworlds (cw). IEEE, pp 284–287

McKinsey & Company (2019) Start me up: where mobility investments are going

Michigan State University (2021) MSU introduces electric autonomous bus, Nov. 8, 2021 [Online]. Available: https://msutoday.msu.edu/news/2021/msu-unveils-new-electric-autonomous-bus

Mobility innovation alliance [Online]. Available: https://mobilityinnovationalliance.org/en/

Riggs W, Schrage N, Shukla S, Mark S (2023) The trip characteristics of a pilot autonomous vehicle rider program: revealing late night service needs and desired increases in service quality, reliability and safety. In: Meyer G, Beiker S (eds) Road vehicle automation. Lecture notes in mobility, vol 10. Springer Nature Switzerland, Cham, pp 93–107. https://doi.org/10.1007/978-3-031-34757-3_9

SIP-ADUS, Strategic Innovation Promotion Program (SIP)—Automated Driving for Universal Services (ADUS) [Online]. Available: https://en.sip-adus.go.jp/sip.

The University of Iowa (2024) Automated Driving Systems (ADS) for Rural America, consulted 18 May 2024 [Online]. Available: https://adsforruralamerica.uiowa.edu/

Workshop on Autonomous Vehicle Data Reporting, California Public Utilities Commission [Online]. Available: https://www.cpuc.ca.gov/events-and-meetings/cped-workshop-2023-06-22

U.S. Department of Commerce, Japan autonomous driving [Online]. Available: https://www.trade.gov/market-intelligence/japan-autonomous-driving